GW00537506

At Home on Planet Earth

An Integrated Course in Science and Religion

Dedication

To the Hon. E. R. H. Wills, founder and Chairman of the Farmington Trust, without whose inspiration and generosity the Science and Religion Project could not have been launched.

At Home on Planet Earth

An Integrated Course in Science and Religion

Derek Sankey, Desmond Sullivan and
Brenda Watson

Basil Blackwell

© 1988 The Farmington Trust Ltd

First published 1988

Published by Basil Blackwell Ltd
108 Cowley Road
Oxford OX4 1JF
England

All rights reserved. No part of this publication
may be reproduced, stored in a retrieval
system, or transmitted, in any form or by any
means, electronic, mechanical, photocopying,
recording or otherwise, without the prior
permission of the publisher.

British Library Cataloguing in Publication Data

Sankey, Derek
    At home on planet earth.
    1. Religion and science——1946–
    I. Title   II. Sullivan, Desmond   III. Watson,
    Brenda
    215   BL241

ISBN 0–631–90048–9

Designed by Bob Prescott

Typeset in Berkeley Old Style series
by The Word Factory (Tonbridge) Limited

Printed in Hong Kong by Wing King Tong Co. Ltd.

Foreword

A recent opinion poll revealed that, among those people who have doubts about the existence of God, one in three said that the reason was that science had 'disproved' religion. Yet we find that a sizeable number of professional scientists are religious believers. For them, science and religion are not seen as alternative views; both are needed if one is to live a fully rounded meaningful life in a scientific age. Having understood through science what the world is like, it then becomes for them natural to go further and to enquire what it all means.

So who is right? *Does* science disprove religion, or are they both necessary for living a full life? Are both necessary if we are to care properly for our home, planet earth? This book sets out the various issues and invites you to work out your own answers to these questions.

Russell Stannard
Professor of Physics, Open University

Acknowledgements

This course has emerged out of the first two years of a four-year science and religion project undertaken by the Farmington Institute. Derek Sankey was the Director of the project for this first part and compiled the main body of the material. It was prepared for publication, with some additions, by Desmond Sullivan and Brenda Watson.

The Farmington Institute is grateful for the help, advice and encouragement given by Martin Rogers, W. P. C. Davies, Dr Peter Hodgson, Derek R. G. Seymour and Richard Wilkins, and to Dr Marjorie Reeves of Oxford.

The authors wish to express their sincere thanks to Mrs B. Colquhoun, Miss S. Jenkins, Mrs P. Stone and Mrs D. Sankey for all their hard work in preparing the typescript.

The authors and publishers would like to thank the following for permission to reproduce photographs:

Barnaby's Picture Library 7.12; BBC Hulton Picture Library 1.11, 2.1b, 2.1d, 2.18a, 2.25a, 2.26c, 3.13, 3.14, 5.4, 8.10; By Courtesy of the British Museum 3.1; California Institute of Technology 7.2; Camera Press 1.13, 2.26d, 8.7; Central Office of Information 3.15; Dominique Berretty/ Colorific! 2.1a; Dover Publications Inc. 8.2; Geological Museum p.35; Sonia Halliday Photographs 5.8, 5.9; Eric Kay 1.6, p.35 bottom; Frank Lane Picture Agency p.37, 3.8, 6.3, 7.11; The Mansell Collection 1.5a, 1.5b, 2.1c, 2.25b, 2.25c, 2.25d, 2.26b, 2.23, 2.27, 3.7, 4.3, 5.10, 8.12; Marion & Tony Morris of South American Pictures 4.7a, 4.7b; John Murray 4.5; NASA 2.9, 5.7, p.89; Oxford Scientific Films Ltd. 4.1a; Palomar Observatory Photograph 7.1; Ann & Bury Peerless 1.4a; Ann Ronan Picture Library 1.7b, 1.8a, 4.6; The Slide File 1.7a; Swift Picture Library 1.4b, 4.1b, 4.2, 6.2, 6.4; John Thompson Animal Photography 4.9a,b,c; Travel Photo International 2.26a; Graham Topping 5.5.

Special thanks to Adrian Harvey for the cover picture.

In addition, the following illustrations have been redrawn:
Fig. 1.8(b) from *Odhams Colour Library of Knowledge, Worlds Beyond Ours*, Aldus Books Ltd.
Table 1, p.40, from *Studies in Christianity and Science: Creation and Evolution*, by Colin Humphreys, published by Oxford University Press. © Colin Humphreys.
Fig. 3.4 from *Genesis: The Origins of Man and the Universe*, by John Gribbin, published by Oxford University Press. © John and Mary Gribbin, 1981.
Diagram p.66, from *Discovering Ecology*, by Tim Shreeve, Sceptre Books Ltd.
Fig. 6.8 from *The Global 2000 Report to the President, Entering the 21st Century*, Viking Penguin Inc.
Fig. 7.15 from *The Times* 22.2.85, Light years ahead of the world, illustration by John Grimwade. © Times Newspapers Ltd., 1985.

Every effort has been made to contact copyright holders. We apologise for any inadvertent omissions and will be happy to make corrections when reprinting.

Contents

Introduction

This course centres on the relationship between science and religion, but it also relates to other areas of the curriculum, notably English and history, and would be suitable for integrated work bringing various disciplines together.

Many people today think that science and religion are opposed. 'The world was made by science and not by God' is the kind of comment that children often make, and it shows a serious misunderstanding of both science and religion. It is hoped that this course may help to unravel the confusion behind such popular statements.

In this book we have tried to help young people to think creatively and imaginatively about such issues; it is not intended to give 'correct' answers. The purpose is not to teach individuals *what* to think, but rather *how* to think. This book does not teach science but how to think about science. Similarly, it does not teach what to believe in religion but how people have believed and do believe now in a scientific age.

The title, *At Home on Planet Earth*, refers to the underlying theme of 'at-home-ness'. Thoughout human history there has been a strong desire to organise knowledge into a cohesive framework, enabling people to feel a sense of being at home in the world. This was one of the main aims of ancient cosmologies. What concept of a unified world view is open to us today?

This raises a number of questions such as 'Have the discoveries of science helped us to feel at home in the world?' and 'Should religion give a sense of "at-home-ness" or does it teach that earth is only a temporary home, and life a journey (as in Bunyan's *Pilgrim's Progress*) to another and eternal world?'

Can modern scientific and religious ideas work together or are they different approaches to the world and opposed to each other? The idea that science and religion *can both* help people to feel at home is developed but in an open way which invites pupils to draw their own conclusions.

The purpose of the story is to base the discussion of these questions and topic work on concrete experience. Quite complicated ideas can be communicated even to young children and certainly to those of secondary school age, if they are put in terms of actual people and events.

Although the story is largely fictional, all the characters – both staff and children – are modelled on people who actually exist and who have taken part in such a camping expedition. It is hoped that this practical grounding will give the characters and indeed the camp itself a touch of realism. The school concerned was a typical comprehensive school with a full range of abilities. The three teachers, one representing religious education, one English and one science, were chosen because the issues which emerge on the course relate most directly to their disciplines within the school.

The topic work is presented in such a way that pupils could manage it on their own. However, the teacher may feel that it would be done more satisfactorily in class.

_____ *1* _____

A sunset experience

Announcing a school camp

*I*T was a bleak March morning at Hillcrest School. The lower school pupils were huddled together in the hall. The look on Wayne's face was typical of many. It bore a simple message: 'I hate school'.

Suddenly the mood changed. Mr Jones, head of English, announced that there was to be a summer camp. 'The plan is to stay at a farm in the countryside. We'll take twenty pupils in the two minibuses, and Mr Payne will be the other driver. (Mr Payne was head of lower school and an RE teacher.) There will be plenty to do. We shall spend a day on the farm and, if we get a clear night sky, we'll learn about astronomy. If the weather is good in the evenings, we can sit and chat around the camp-fire. On one of the days there'll be a biology field study. To help with that, Miss Ridgewell will be joining us.'

That caused a murmur of chatter. Miss Ridgewell was the new science teacher and, although she was rather strict in class, she was also popular. Mr Jones raised his voice above the growing noise. 'Now, if you want to come to camp, you will need to fill in the form which is at the back of the hall. Get your parents to sign it and to return it quickly. It'll be first come, first served.'

Setting up camp

The places were soon filled, but it seemed ages before the day of the camp came round. When it did finally arrive, the campers were relieved to see that the rain, which had fallen steadily for the previous week, had stopped.

'We might be in for a week of good weather,' said Mr Jones as he loaded the last of the luggage into the minibuses. By the time they were on the road the sun broke through. The countryside took on a new appearance as everything outside brightened up.

The field where they were to make camp was partly surrounded by trees; so there was plenty of fallen wood for making a camp-fire. While

the teachers were organising the food, everyone else went into the woods to explore and to gather firewood. Some of them returned quickly, unwilling to venture too far. Not Wayne, though – he went deeper into the wood and he came back last of all. 'I reckon I know more about this place now,' he said, full of confidence. 'So it feels a bit more like home – it's good here, Sir.'

Dawn, who hated woods and trees, was not so sure: 'It doesn't feel much like home to me, not yet,' she said. 'Just think what it will be like with those trees when it gets dark. There won't be any street lights you know. It'll be creepy.'

Sunset

It was getting late. The campers sat in the open, drinking hot chocolate with their supper. At the end of the field, where there were no trees, the sun was beginning to set. Gradually it dipped towards a ridge of hills in

the distance, seeming to grow ever larger. Its colour changed to the deepest red. All around, the evening sky began to burn in the radiance of the sun's reddening glow. Higher in the sky the large mountainous clouds were flecked with gold.

Everyone stood still, captured for a moment by the magic of what they were seeing. It was like a parting farewell. The sun gradually sank below the horizon and finally passed out of view.

'That was fantastic,' said Leila, as everyone returned to what they had been doing before. 'I've never seen a sunset quite like that.' 'I have,' retorted Wayne, 'and anyway its not that wonderful; it's only the earth turning on its axis.'

Mr Payne, who had stood enchanted a little longer than the others, picked up Wayne's comment. 'Is that *all*, Wayne? Don't you think perhaps you're missing something?'

(The story continues on p. 15.)

Things to do

Thinking about the story

1　Do you think that Wayne was 'missing something'?
　　If so, write down in one or two sentences what you think he was missing.

2　Read carefully the following three passages about the sun.

　　a　The sun is a very ordinary star and, compared with many other stars in our galaxy, it is neither very large nor very bright. The reason why it looks large and bright to us is that it is very much nearer than any other stars. The earth is one of the sun's planets. Every 24 hours the earth turns once on its axis. The side of the earth which faces the sun experiences daytime, whereas the side which is turned away is in night. What we experience as sunset is the result of our part of the earth moving away from the sun's light. The sun gets closer to the horizon and, as it does so, it seems to grow larger and to change colour from yellow to red.

　　b　Full-blooded, red against the western sky,
　　　　The mighty warrior prepared to die,
　　　　Gently reclining his wounded head
　　　　Upon his pillowed, hilly bed.
　　　　And all the world was heard to say,
　　　　'Will he rise again – one day?'

　　c　Day is dying in the west,
　　　　Heaven is touching earth with rest;
　　　　Wait and worship, while the night
　　　　Sets her evening lamps alight
　　　　Through all the sky.
　　　　Holy, holy, holy, Lord God of hosts:
　　　　Heaven and earth are full of Thee,
　　　　Heaven and earth are praising Thee,
　　　　O Lord most High.

What kind of book do you think each passage might have been taken from? Is one more true than the others? Which describes best what we see with our eyes? Which passage would Wayne prefer? Which passage do you prefer, and why?

Creative responses
1   Describe in words, or in a picture, any experience that you have had which made you want to stand still, like the people in the story.

2   Design a poster showing the sun and a variety of living things on earth which depend on the sun for their life.

Topic work: Ways of seeing

1 Seeing is more than using our eyes

Have you ever found yourself in a room when the lights have gone out and it is pitch dark? How can you find your way when you cannot *see* where you are? Perhaps you have played the party game when things are put into a box and you have to work out what they are without seeing them. Why is this often difficult when you can only use your sense of touch?

Look at these pictures (Figure 1). Try to stare hard for some time at the picture of the steps (Figure 1(c)).

into giving the wrong answer, you are the victim of an optical illusion.

When we try to identify things in the dark by touching them, it is obvious that we are using our brains as well as our hands. We have to use our *imaginations* and we remember what things that we have known *feel like* and *look like*. When we are using our eyes to see, it is not always obvious that we are using our brains as well. However, these kinds of pictures seem to show that our brains do influence what we see – even the way we see a sunset (Figure 2).

(a) An old witch or a beautiful young lady?

(b) Which line is the longest?

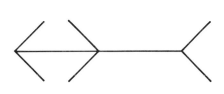

(c) Stairs which never end

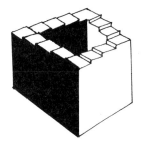

(d) Which of these is larger?

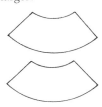

(e) Stairs seen from above? or from below?

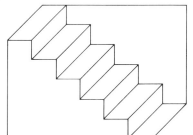

(f) Which girl is the tallest?

Figure 1 Optical illusions

What do you notice? We usually call pictures like these **optical illusions**. They are fun, but they can also help us to understand that there is more to seeing than using our eyes. When we see something, the light from an object makes an image on the retina of our eyes. In other words, our eyes show us what the object is like. But, if it is as simple as that, why can we 'see' the same picture in different ways? Look at Figure 1(d), for example. Is it just a pattern of lines, or stairs going up, or stairs coming down? The image of Figure 1(d) is formed on the retina of your eye, but your brain interprets this image and 'tells' you what the picture is. If your brain is 'tricked'

Figure 2 As the sun sets, it seems to grow larger

There are other ways of 'seeing' too, quite different from looking at a sunset. Study these three sentences.

1 'The judge *sees* through him all right.'
2 'Sarah *sees* the mouse in the corner and screams.'
3 'Jonathan *sees* the world as a hostile place.'

Only one of these 'sees' means 'seeing with physical eyes'. Which one? The other two 'sees' mean seeing with the mind or imagination. Write down the meaning of the two sentences without using the word 'sees' or any word very similar to it. In these two sentences, 'sees' is used as a **metaphor**. A **metaphor** describes one thing in terms of another.

2 *Asking questions about the world*

Since the day we were born we have been learning more and more about the world in which we live. Have you ever noticed how very young children begin to ask searching questions almost as soon as they can talk?

Being able to speak marks a very important moment in our growing up, but we started to explore the world long before we were able to ask questions. By asking questions, we gradually build up a picture of the world around us. At first we tend to rely on the answers given by our parents and teachers, but, as we grow into adults, we begin to make up our own minds and to have our own ideas.

Marianne, who was 12 years old, was talking to her father over breakfast. 'I was thinking the other day', she said, 'how strange it is that we are living on a planet in space. When I was younger, I thought that the earth was solid – on the ground, so to speak – but then it suddenly struck me that the earth is moving around out in space. Perhaps we are the only people in the whole universe. It's really weird why we are here.' 'I can't really explain what I mean,' she added, as she shrugged her shoulders, 'but I suddenly felt how strange it all is. I'd never thought about the world like that before.'

Have you ever felt that the world was strange like that? Make a list of questions and ideas that come to your mind when you think about the natural world and about life on earth. Here are some examples.

1 Do you sometimes wonder what caused the world in the first place, and whether it was made for a purpose?
2 Do you feel 'at home' in the world? Does it sometimes seem unfriendly and frightening?
3 Have you ever wondered how we human beings fit in with the rest of nature? Are we different from other living things? Are we similar to other animals or plants or even to the stars in some way?
4 Are you sometimes struck by the wonder and beauty of the world, for instance when you see a gorgeous sunset or some splendid scenery?

Talk these questions over in class or, if you are working on your own, think them through carefully yourself. Can you write down similar questions of your own about the world and life on earth?

Figure 3 shows three different ways of looking at the world. See if you can match these three pictures with each of the questions 1–4 above. You may decide that some of the questions can be looked at in more than one way.

The children in the story lived in a large city and, for many of them, it was a new experience to spend a week in the country. It is sometimes said that our

Figure 3 Different ways of approaching the world

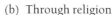

(a) Through art   (b) Through religion   (c) Through science

(a) A Hindu cremation

(b) A bird on its nest

Figure 4 Experiences that make people think more deeply about life

modern way of life separates us from the world of nature.

- Do you think that this is so?
- Do the seasons mean much to you?
- Do you think it is true that poeple who live in the countryside enjoy nature more than people who live in the city?
- Make a list of situations and events which help people to think about the world and their lives on this planet. For clues, see Figures 4(a) and 4(b).

3 Ways of seeing the world: stories from the past

Throughout history, people have asked questions about the kind of world that we live in, such as whether it makes sense, or whether we can feel at home in it. They often realised how much their lives depended on the sun. Many early peoples therefore believed the sun to be a god (Figures 5(a) and 5(b)).

In ancient Egypt, people worshipped many gods and goddesses including the sun.

Figure 5(a) Inti, the sun god of the Incas of Peru

Figure 5(b) Apollo, the sun god of ancient Greece

Figure 6 is a picture of a stone carving from ancient Egypt. It is almost 3000 years old. The figure in the centre is the sun god Re. On either side are Pharaoh Akhenaton and his wife Nefertiti. In their hands they hold sacrifices which they are offering to the sun. In return, the sun is blessing them with the gift of life.

Some worshippers probably did not worship the sun itself but the 'power behind the sun'. The sun was really a *symbol* of the god, but it was easy for people to confuse the two.

Akhenaton saw this very clearly and he gave the sun god another name, Aton, to remind himself and his people to worship the power behind the sun. Indeed, he believed this power to be the only god there is.

Figure 6 The sun god of ancient Egypt, Ra or Re

The Babylonians were keen astronomers and made some very accurate observations of the stars. They were able to forecast eclipses of the moon. Their measurement of time was very precise. They divided hours into 60 minutes and minutes into 60 seconds.

Yet, like all the other ancient civilisations, their view of the sky was mainly religious. The heavens were thought to be the home of the gods, and the sun and planets were thought to be gods. They developed the signs of the zodiac and studied the movement of the stars in order to learn how they influence human life (astrology). However, there were other practical reasons for knowing the positions of the sun and the stars.

● What is the connection between the pictures in Figure 7?
● What do we mean by 'a day', 'a year' and 'a season'?
● How would a knowledge of the sun's position help (i) farmers and (ii) priests? See whether you can find out how the date of the Christian festival of Easter is fixed each year and why the date changes.

Science

The beginnings of modern science are to be found in ancient Greece. In Athens, around 2500 years ago, certain people started to ask questions about the universe. They were not content simply to record the movements of the sun and planets; they wanted an *explanation* or *theory* of how and why they moved.

(a) Sunset seen through a dolmen on The Burren,
County Clare, Eire

Figure 7

(b) Sundials are often found at the doors of Christian
churches

One theory suggested by Aristotle, the teacher of
Alexander the Great, influenced the way in which
people were to think about the world for nearly 2000
years. Two interpretations of Aristotle's theory are
given in Figure 8. For fuller details of Aristotle's
theory, see Unit 2.

Figure 8(a) shows a mediaeval adaptation of
Aristotle's theory of the universe. See whether you
can work out the order of the spheres. The names are
given in mediaeval Latin. Lunae is the moon. Solis is
the sun. The outer ring is the 'habitaculum (home)
dei (of God)'. The others you can probably guess.

Aristotle believed that the earth is at the centre of
the universe and that the planets and the stars move
around the earth on invisible spheres. Surrounding
the sphere of the earth is another sphere (that of the
moon), then another (that of the planet Mercury)
and so on until we reach the outermost sphere which
drives them all round. That is the sphere which
Aristotle called the First Mover. For Aristotle, the
First Mover started the whole universe in motion.

Most of these early scientists saw no conflict
between their scientific discoveries and their
religious beliefs. Aristotle, for example, did not
believe in the gods of Mount Olympus such as Zeus
and Aphrodite (whom the Romans called Jupiter and
Venus), but he believed in the First Mover.

The mathematician Pythagoras, whose famous
theorem (Figure 9) is still learnt in school, sacrificed
an ox in a temple as a thanksgiving to the gods for
helping him to work out the theorem.

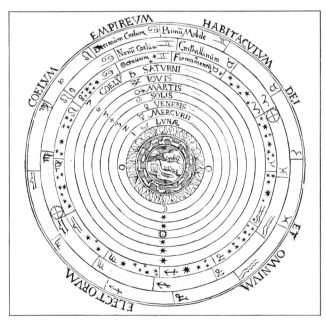

(a) A mediaeval picture of Aristotle's theory of the
universe

Figure 8

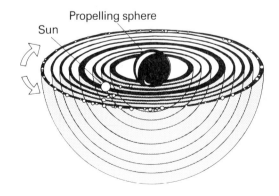

(b) A modern attempt to picture Aristotle's world of
spheres within spheres

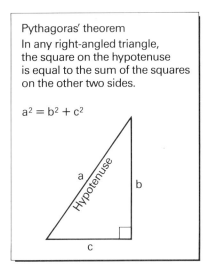

Pythagoras' theorem
In any right-angled triangle,
the square on the hypotenuse
is equal to the sum of the squares
on the other two sides.

$$a^2 = b^2 + c^2$$

Figure 9
Pythagoras' theorem

We no longer believe Aristotle's theory of the universe. However, he and other Greek philosophers made an important step forward in trying to find out how the world works. This search for better and better scientific explanations of the world has continued.

Does our quest for scientific knowledge mean that we no longer need to approach the world through religion or to notice the beauty and wonder of the world? A modern philosopher said: 'When you understand all about the sun, all about the atmosphere and all about the rotation of the earth, you may still miss the radiance of the sunset.'

What do you think he meant by saying that? Do you agree with what he is saying?

4 Ways of seeing science and religion: a look at a text

In the ancient world, science and religion were mixed up. But today many people see them as distinct. The different ways in which people see the following passage from Genesis reveal the different ways in which they see science and religion.

Genesis 1, 1–8

1 In the beginning God <u>created</u> the heavens and the earth.
2 And the earth was empty and desolate and there was darkness over the waters, and the <u>spirit</u> of God <u>moved</u> upon the waters.
3 And God <u>said</u> let there be light: and there was light.
4 And God <u>saw</u> the light, that it was good: and God <u>divided</u> the light from the darkness.

5 And God <u>called</u> the light Day, and the darkness Night. And there was <u>evening</u> and there was <u>morning</u>, <u>one day</u>.
6 And God <u>said</u> let there be a <u>dome</u> in the middle of the waters, and let it divide the waters in two.
7 And God <u>made</u> the <u>dome</u>, and <u>divided</u> the waters which were below from the waters which were above. And it was so.
8 And God <u>called</u> the <u>dome</u> Sky. And there was <u>evening</u> and there was <u>morning</u>, a <u>second day</u>.

(Look up the rest of Genesis 1, and Genesis 2, 1–4, in a Bible.)

Read the passage carefully. What do you think of it?

● Do you think that science shows that this passage is wrong?
● Do you think that this passage shows that science is wrong?
● Do you think that this is a story like Aesop's Fables or like parables? If so, what do you think its message is?
● Do you agree with the story in Genesis 1 and Genesis 2, 1–4? If so, do you see it as literally true or as poetically true?
● Have you any other ideas about it?

In science, if someone thinks that the earth is flat, they are wrong. In matters relating to religion, it is more difficult to show what is right. People who have studied religion have very different opinions. Figure 10 is a chart which shows this.

It is important that our opinions are based on understanding and not on ignorance or blindness. In looking at a passage like this one from Genesis, we need to remember three 'rules'. Otherwise we may make some bad mistakes.

Rule 1 Try to understand the framework or *context* in which the passage was written. 'Context' is a useful word indicating that something does not stand on its own but is part of something bigger. For example, 'the earth must be studied within the *context* of the vast system of galaxies.'

Rule 2 Try to understand what the words mean in the original language in which they were written. It is difficult to translate words adequately from one language to another.

Rule 3 Try to understand how the words are used. Are they used literally or as metaphors?

Let us apply these three rules to the passage from Genesis.

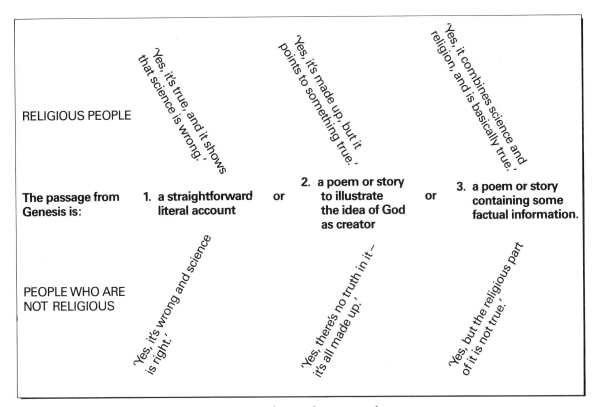

Figure 10 Chart showing the ideas of religious and non-religious people

RELIGIOUS PEOPLE

'Yes, it's true, and it shows that science is wrong.'

'Yes, it's made up, but it points to something true.'

'Yes, it combines science and religion, and is basically true.'

The passage from Genesis is:

1. **a straightforward literal account** or 2. **a poem or story to illustrate the idea of God as creator** or 3. **a poem or story containing some factual information.**

PEOPLE WHO ARE NOT RELIGIOUS

'Yes, it's wrong and science is right.'

'Yes, there's no truth in it – it's all made up.'

'Yes, but the religious part of it is not true.'

Rule 1 *The context*

The passage from Genesis was first written down about 2400 years ago. But the story was known long before that. It had passed down by word of mouth from one generation to another over hundreds of years. The authors (almost certainly *many* people contributed to it) were Hebrews, ancestors of present-day Jews. They lived an open-air life, first as nomads and then as farmers and traders in Palestine (Figure 11). They were expert in their knowledge of the seasons, and used the sun and stars for finding their way. They were also great story-tellers and loved thinking about the many stories, ideas and information which they heard on their travels throughout Syria and Mesopotamia (Figure 12). The tales which the Hebrew people told and wrote down relied on the ideas of other people. But, far from copying ideas and stories that they had heard, the Hebrews used them in a very original way.

The passage from Genesis was regarded by them as so important that it was put right at the beginning of their scripture. It may have been part of a service for worship in the Temple at Jerusalem (Figure 13).

Figure 11 The Hebrew people were originally nomads. Today, the Bedouin still live a nomadic life

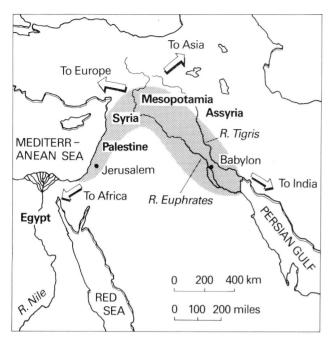

Figure 12 The area covering Palestine, Syria, Mesopotamia and Assyria was known as the Fertile Crescent

Figure 13 The Wailing Wall is all that is left of the great Temple at Jerusalem

Rule 2 *The words themselves*

The passage was originally written in Hebrew. The first sentence is shown in Figure 14 (Hebrew is written from right to left).

Scholars find the study of words fascinating. For example,

rē'sh (which is part of the first word) began by meaning just 'head', but later was used for 'at the head of,' 'the top of' and 'the start of'. So it came to mean 'begin'. Here it means 'the beginning of time'. If we identify this point as the beginning of time, then we cannot use the word 'time' to describe any period which went before it. Before the beginning of time, we cannot talk of 'time' or 'no time'. This is an

extremely difficult idea for us to understand – as difficult as the idea of the earth being 26 billion miles away from the nearest star. Even if we cannot understand the idea fully, we can at least realise that we cannot understand it and simply accept it, as we do with the billions of miles.

bārā is an extremely important word which we should look closely at. It is usually translated as 'create' or 'make', but in Hebrew it is *not* the usual word for making bread or making a pot or creating a work of art; that is **yatser.** Bārā is a word which the Hebrews tried to keep special. It is used sixty-one times in the Jewish scripture but always only of God's creating the world.

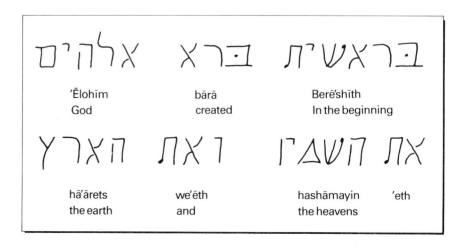

Figure 14 The first sentence of the book of Genesis

Rule 3 How the words are used

The words underlined in the passage on page 10 are all **metaphorical.** The Hebrews found it impossible to speak much about God without using language in this way. Religious people have always had to do this. Otherwise they would lapse into silence because God is infinitely greater than anything they can possibly speak about. So they talk about God *as though* 'He' were human or a part of this world. 'He' is an example of metaphor. If people think of God as 'She' or as 'It' that too is metaphor.

It is important to understand this use of language because otherwise silly mistakes can be made. The sentence below uses a metaphor. Explain exactly why both comments, A and B, are stupid.

'The man is an ass'

A: 'No he isn't because he hasn't got long ears.'

B: 'So therefore he's got long ears.'

Do the same for this sentence:

'And God saw the light that it was good.'

A: 'No he didn't because he hasn't got eyes.'

B: 'So therefore he has got eyes and stood and looked at what he'd created.'

Figure 15 'The man is an ass'

And these two:

'And there was evening, and there was morning: that was one day.'

A: 'No! The earth evolved over billions of years not in twenty-four hours.'

B: 'So therefore it all happened at once, in the space of twenty-four hours.'

2
Can we be sure of what we know?

An accident

After an early breakfast, Wayne, Dawn and Carol didn't like the idea of washing up, and so they offered to collect the milk from the farm. It was some time before they returned and the teachers were beginning to wonder if they had lost their way. When at last they did arrive, they seemed very excited.

'There's been an accident with a milkfloat and a car,' shouted Wayne, as they neared the camp.

'Did you see what happened?' asked Mr Payne.

'Yes,' says Wayne, 'there's milk all . . . .'

'No, he didn't,' said Dawn. 'It must have happened just before we got there.'

'But I heard it, anyway,' replied Wayne.

'No, you didn't. He's telling lies,' said Carol.

'No, I'm not, I can prove . . . .'

'Just a minute,' interrupted Mr Payne, 'you three will finish up in a fight if you're not careful. Just tell us whether anyone was hurt.'

'The milkman had a bad cut on his head,' said Wayne.

'It wasn't *that* bad,' replied Carol.

Mr Jones had been listening nearby. 'Are you sure you all saw the same accident?' He asked.

'Yes we did,' replied Wayne, 'but as usual the girls have got it wrong, and it was the woman's fault too . . . the accident.'

'No it wasn't,' said Dawn. 'It was the stupid milkman. He had left the milk van on a corner and she couldn't see it. The trouble with Wayne is that he is biased. He's always like that. He thinks that all women are useless drivers and men never do anything wrong.'

Miss Ridgewell, meanwhile, had come across because of all the noise. She had heard enough. 'Surely we are not going to spend the rest of the day arguing about what happened!' she butted in.

'I hope not,' replied Mr Payne, 'but it just shows you that three people can see the same thing and yet see it differently. I know, let's talk about it again tonight. Then all of you can tell us how it really happened.'

Around the camp-fire

Around the fire that night Carol, Wayne and Dawn each gave their own story of how the accident had happened, but they were still unable to agree on the details.

'Why can't you three agree? You all saw the same thing, didn't you?' Mr Payne asked.

'Because Wayne isn't telling the truth,' said Dawn.

Before Wayne could say anything, Mr Payne quickly interrupted them: 'I don't think that's the real reason. Nor does Wayne, I am sure. So, before you two start arguing again, has anybody else got the real reason?'

'*We* don't know who's telling the truth, do we?' protested Winston. 'We weren't there and so we can't tell what happened. It's the same problem with newspapers, isn't it? They often give different stories of the same events.'

'Yes, but why? I mean, all you've got to do is to use your eyes,' added one of the girls.

'Isn't that the trouble?' suggested Mr Payne. 'We may see things with our eyes, but we ourselves have to make sense of what we see. So each of us might see things differently.'

'But we can't choose to see what's not there,' chipped in Wayne.

'Yes, you can, and that's why you got it wrong, Wayne,' replied Dawn. 'You just saw what you wanted to. You think that women are bad drivers but you've got it wrong.'

'So how do we know who is right, then?' he replied. 'It's like science, for instance. That's based on what we see with our eyes and we can prove it's true,' Wayne said, looking straight at Miss Ridgewell.

'Who told you that, Wayne?' she exclaimed. 'I'm certain I didn't.'

'What do you mean?' said Wayne, rather taken by surprise.

'Well, think about it for a moment,' she added. 'Last night we saw the sun go down below the horizon and the stars move across the sky. That's what we see with our eyes. Our common sense tells us that the sun and stars go round the earth. Because of science, *we* know that this is wrong, but not everyone wanted to change their ideas.'

'That's right. The discovery that the earth goes around the sun caused quite a lot of argument,' added Mr Payne.

'I see what you're getting at – science isn't just common sense, or based on what we see. It even sometimes goes against common sense. So how can we be certain of anything?' said Wayne.

A game of certainty

'Many people get angry when we tell them they are wrong. A Frenchman called Descartes was very worried by all this uncertainty and so he played a sort of certainty game. We can try it now, if you like,' said Mr Payne.

'All right, then,' said Wayne, 'what do we have to do?'

'Well, the idea is that everyone thinks of one thing which they are absolutely certain of. All the others try to show that he or she is wrong,' said Mr Payne.

'That's easy,' said Wayne. 'I'll start. I am certain that the sun is shining right now. Nobody can doubt that.'

'I can', replied Dawn, 'because we can't see the sun. It's almost dark.'

'Yes, but it is still shining somewhere. *We* can't see it, but in Australia they can,' replied Wayne, indignantly.

Miss Ridgewell took up Wayne's challenge. 'How can you be certain that Australia exists? You've never seen it,' she said. 'Besides even if you can actually see the sun, you can't be certain that it is *still* shining. The light from the sun takes 8 minutes to reach us, and so we can only say that it *was* shining 8 minutes ago. If the sun exploded, it would take 8 minutes before we knew about it'.

Wayne looked perplexed. He hadn't expected to be beaten *that* easily.

As they passed around the circle, many different suggestions were made. Carol thought she could be sure that they were all at camp and sitting around the fire, but one of the boys suggested that they could all be dreaming.

'I still think that in science we can prove things are true,' argued Winston.

'That's what a lot of people think,' replied Miss Ridgewell. 'Scientists usually think that they are getting closer to the truth, but there is always the possibility that today's theories will turn out to be only partly true. History shows that scientific knowledge changes.'

'So what about this Frenchman? Did he ever win his game? Did he find out something certain?' asked one of the girls.

'Yes,' said Mr Payne, 'the thing that he believed he could be certain about was himself,' replied Mr Payne.

'What do you mean by that?' asked Winston.

'Well, Descartes knew he was thinking, didn't he? And, if he was thinking, then he believed that he must exist or he wouldn't be able to think.'

That set everybody thinking, trying to see whether they could find an argument against their own existence.

It seemed as though the discussion could go on all night, but some felt it was getting late and the fire started to lose its warmth in the cold night air.

Things to do

Thinking about the story

1 You may like to try playing 'the game of certainty' in your own class. Begin by having a few minutes of silence while you jot down on a piece of rough paper anything that you think you can be certain about. Try to think of something which was not mentioned in the story.

2 Can you find some arguments against Descartes' view that you can be certain of one thing: that you exist?

3 'If we waited until we were certain, we would never do anything!' Is this good advice to a football team? Is it good advice to an athlete?

4 Why do you think that we often see things differently from other people? Sometimes you may see examples of this on television when 'eyewitnesses' seem to be giving different accounts of the same incident.

5 How often in life do you have to rely on what other people tell you? Can you think of ways in which we are *bound* to rely on what we are told both in science and in religion? Why do you trust some people and not others? Is it because they can always *prove* what they say or are there other reasons?

6 What did Miss Ridgewell mean when she said that the things we know in science sometimes turn out to be only partly true?

Creative responses

Either

1 Write your own short story (about 200 words) describing a discussion on the telephone between two penfriends from different countries who have just watched an international sporting match on television between their two countries. One friend saw that his or her country did better than the other. Try to show in your story how, in a way, they might have seen some of the incidents differently from each other.

Or

2   Work in pairs and together write a short piece of conversation
    between two opponents who are arguing about the truth of
    something in the news recently using the ideas of 'proof', 'certainty'
    and 'biased'.

Topic work: Making sense of things

1 Ways of knowing

The story drew attention to a very large problem – how can we know anything? Can we be sure that what we think we know is really true?

There are basically four ways in which we can come to know anything.

1   By seeing something physically with our eyes or by using our other four senses of hearing, smelling, touching and tasting.
2   By 'seeing' something with our 'inner' eyes, i.e. by how we grasp things, what we somehow 'just know' even though we cannot see it with our physical eyes, as when we say 'the penny drops' or after a joke 'Do you get it? Do you see the joke?'
3   By using our reason, and thinking something out.
4   By accepting what other people have seen or thought out, as in books or news broadcasts, or listening to what people say in conversation.

In Unit 1 the campers all saw the sunset with their physical eyes. Most of them 'saw' a great deal more with their 'inner' eyes. In the work topic in Unit 1 about ways of seeing, you were using your reason and, if you agreed with the scientific description of the sunset, you were accepting what other people have worked out.

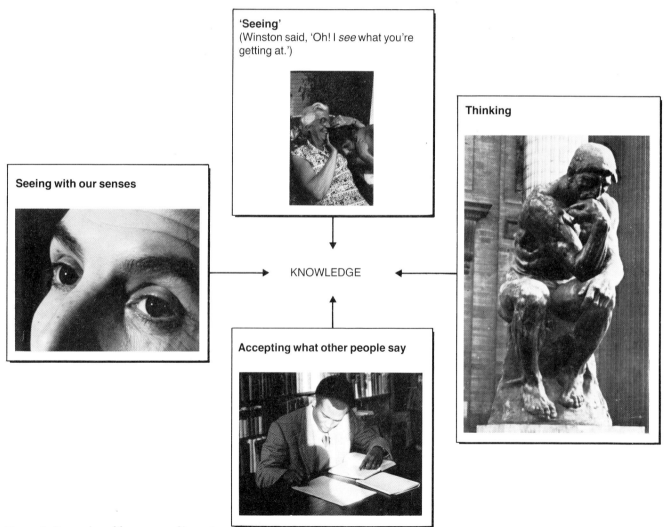

Seeing with our senses

'Seeing'
(Winston said, 'Oh! I *see* what you're getting at.')

Thinking

Accepting what other people say

KNOWLEDGE

Figure 1 Examples of four ways of knowing

These different ways of knowing have been given names.

1   What we know through using our senses is called everyday knowledge or *empirical* knowledge.
2   What we know through using our feelings is called inside knowledge or *intuitive* knowledge (from the word intuition).
3   What we know through thinking is called *rational* knowledge (reason).
4   What we know through accepting what other people say is not often given a special name, but we can call it *authoritative* knowledge because we accept it on someone else's word or authority.

Read again the section of the story called 'Around the camp-fire'. The campers give examples of all four ways of knowing. Copy the chart in Figure 1 and write an example in each box. One has been done for you.

Copy the chart in Figure 2 and place each of the following four words in the appropriate box.

Intuitive
Authoritative   } knowledge
Rational
Empirical

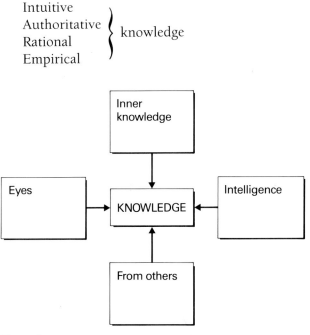

Figure 2

Who was right in the argument about the accident? Did both Wayne and Dawn really *know* what happened or did they only just *think* that they knew?

2 Who is right?

Wayne and Dawn could not both have been right. Whoever was in fact right could not prove it. The question of who is right is something which people have argued about throughout history. Something which everybody asks is: 'How can I be sure?' Scientists especially are concerned with this kind of knowledge. They try to reason things out, but each scientist does not start from scratch, working everything out alone. Scientists begin by accepting what other scientists have taught. They add new knowledge from their own research, which is then tested and thought about by other scientists.

In Section 3 we shall see that scientists do not rely on any one way of knowing. They use **authoritative knowledge** (believing other scientists), **empirical knowledge** (from their own research), **intuitive knowledge** (from their own experience) and **rational knowledge** (from their own thoughts). Each way of knowing has to be balanced and corrected by the other ways. This has led to development and change in the way in which scientist see the world. When you have read through this section of Unit 2, copy and complete Table 1.

Table 1   Famous scientists

	Ptolemy	Copernicus	Galileo	Newton
What did he see?				
What did he 'see'?				
What did he 'think out'?				
What did he accept from others?				

Religious people are especially concerned with intuitive knowledge but they also take into account everyday knowledge and reasoning. Just as scientists are part of a community in which people share what they have found out, so religious people are almost always linked with a religious tradition. Even if they do not go to church or attend a synagogue, mosque or other religious building, they are usually influenced by the thinking of religions such as Judaism, Christianity, Islam, Hinduism and Sikhism.

As in science, so it is in religion; people have to be very careful not to make wrong conclusions by trusting just one way of knowing. Everything needs to be weighed up against other ways of knowing, and people need to keep on learning and correcting (Figure 3).

Figure 3 Maintaining the balance

3 Seeing the truth about the stars

- When we look at the sun, moon and stars from our position on earth, what do we actually see?
- Do we see all the sky? Or only half? Or less?
- Can we see the earth moving through space?
- Can we see the sun, moon and stars moving around the earth with the earth standing still? (Look at the moon on a windy night as the clouds race by.)
- Do we see the sun climbing up in the sky in the morning, and going down in the evening?

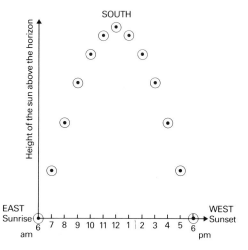

Figure 4 The sun's apparent path across the sky

Figure 4 shows the path of the sun across the sky when it rises at 6 am and sets at 6 pm. Every day the sun *appears* to make a similar journey. We first see the sun in the eastern part of the sky and, by the time that it 'sets' in the evening, it has moved across the sky towards the west. After sunset, we do not see it again until it appears in the east on the following morning. The picture also shows something more about the movement of the sun.

- What line or shape does the sun appear to trace on its journey across the sky?
- What do you notice in Figure 4 about the position of the sun at midday?

During the day we cannot see the stars because of the bright light of the sun. After the sun has 'set', at twilight, the brightest stars begin to appear. As the night grows darker, more and more stars come into view.

At night, the stars appear to move from east to west – like the sun during the day. People who live north of the earth's equator see the stars as circling around a point in the sky directly above the North Pole (Figure 5). This is close to a star which we call the Pole Star or Polaris. South of the equator, people see the stars moving in arcs around a point directly above the South Pole.

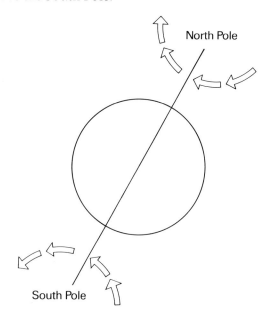

Figure 5 Stars circulating the North and South Poles

Here is a simple observation for you to make yourself. Use a compass to find the direction of north; then look up at the sky and see whether you

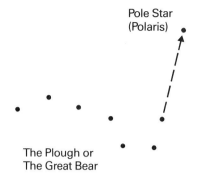

Pole Star
(Polaris)

The Plough or
The Great Bear

Figure 6

can find Polaris by using the group of stars which we call the Plough (Figure 6). These stars also have a shape which is rather like a saucepan with a bent handle and hence in France they are called the casserole.

If you do this early in the evening and then look again some hours later, you will see that the stars in the Plough have moved around but that the Pole Star has remained in the same position (Figure 7).

For thousands of years it was believed that the earth was at the centre of the universe and that the sun, stars and planets really did circle round the earth. This earth-centred theory of the universe seems to fit the facts as we see them. It seems like common sense, but the same facts can be seen in another way, from another point of view. Maybe there were other reasons which made them *want* to believe that the earth was at the centre of things. Can you think what they might have been? (For example: did it make them feel important?)

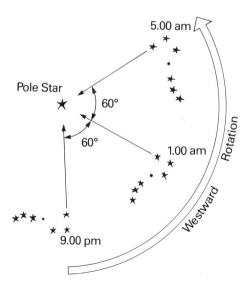

Figure 7 The changing position of the Plough on a night in late October

4 *The problems of the planets*

The wanderers in the night sky
Around 2500 years ago, Aristotle produced a theory of how the universe works (see Unit 1, Topic work). In many ways his theory fits in well with what we see. It also gives a compact and homely picture of the world, with the earth in the centre and the sun, planets and stars moving majestically around the earth on invisible crystal spheres.

Aristotle (384–322 BC) produced a *theory* of what the universe is and how it works.

What is it?	*How does it work?*
All heavenly bodies are made of 'quintessence' – an eternal substance, more noble than any earthly thing. Arranged in spheres, they increase in perfection the farther they are from earth. The farthest sphere is the most spiritual and is a First Mover, which causes all the other spheres to move. This description tells us the 'nature' of the universe.	The heavenly bodies move in perfect circles; the outer sphere makes the next sphere move, which in turn makes the next sphere move and so on until the moon is made to move. The earth is stationary, corrupt and subject to change – unlike quintessence. These movements account for the 'appearances' of the heavens.

So Aristotle believed that you could know the *nature* of things by using your mind but that you can know how they behave (their '*appearances*') by using your five senses.

Aristotle's picture, however, does not quite fit a well-known fact, as we shall show.

Take a look at the two pictures in Figure 8. They were drawn from photographs taken at a planetarium, but they show the paths of the planets over a period of months. Figure 8(a) shows the path of planet Mars; Figure 8(b) shows other planets moving through the same part of the sky. Can you describe what the planets seem to be doing in the pictures?

If you were to make constant observations of the planets, you would see them wandering across the

Figure 8 The wandering paths of the planets over a period of months

background of the stars (Figure 9). In fact, the word 'planet' means 'wanderer'. Aristotle's theory of the universe does not take this wandering movement of the planets into account.

When theories do not quite fit the facts

Although scientific theories try to account for all the known facts, there may be some facts which do not fit perfectly (Figure 10). Also, new facts may come to light which cannot be accounted for by the theory. These misfits obviously pose a problem, and scientists have to make a choice. They can

1 say that the theory is mostly right and ignore the facts that don't fit,
2 suggest changes to the theory so that it fits the facts more accurately or
3 say that the present theory is probably wrong and look for a better one.

Making this choice is not easy. There are no rules to say which alternative is best. It has to be a matter of judgement (Figure 11).

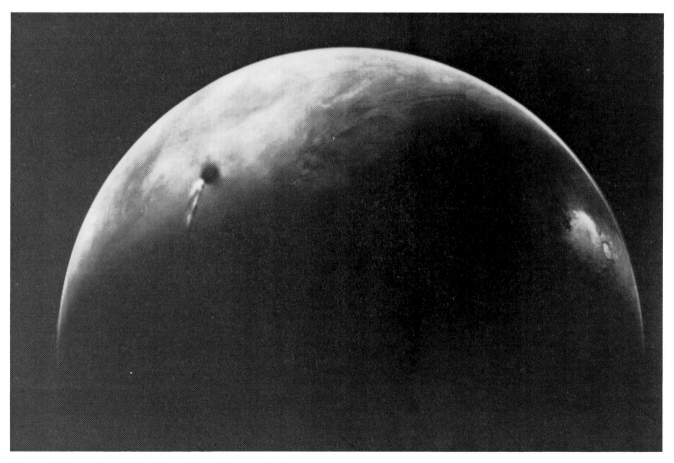

Figure 9 The surface of the planet Mars photographed by a recent probe

Figure 10 When theories don't fit all the facts

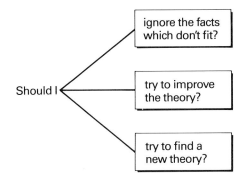

Figure 11 Adjustment of a theory

Ptolemy's theory

500 years after Aristotle, a Greek geographer and astronomer named Claudius Ptolemy (pronounced tol-emi) tried to solve the 'wandering planets' problem. He did it by altering Aristotle's theory using *epicycles*. An epicycle is a circle on a circle as in Figure 12.

Look at Figure 13 which shows the new style of fairground roundabout. In addition to going round like a normal roundabout, the rider's own seat moves in a smaller circle at the same time.

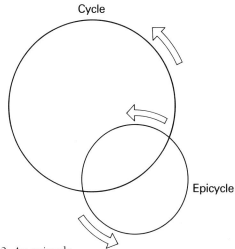

Figure 12 An epicycle

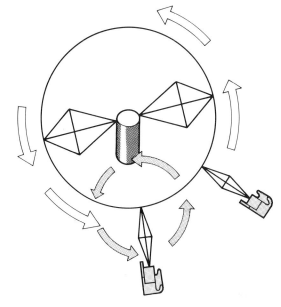

Figure 13 A fairground roundabout

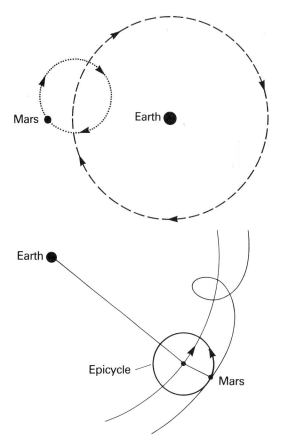

Figure 14 The movement of Mars

The two pictures in Figure 14 show this movement for the planet Mars. Try to make a simple model of this either using a spirograph or with card and string. See whether you can show how Ptolemy's system of epicycles produces the looping effect

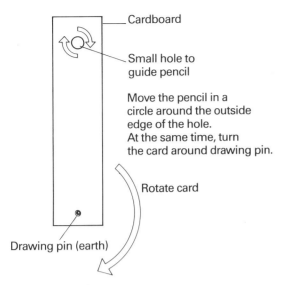

Cardboard

Small hole to guide pencil

Move the pencil in a circle around the outside edge of the hole. At the same time, turn the card around drawing pin.

Rotate card

Drawing pin (earth)

Figure 15 How to draw an epicycle

which we can see that the planets make in the sky. (Cut out the small round disc of the epicycle and fix the centre of the other end of the cardboard to the 'earth' with a drawing pin (Figure 15). Trace the two movements with a pencil (Figure 16).)

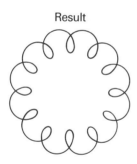

Result

Figure 16 An epicycle drawn by the method shown in Figure 15

By using the idea of epicycles it was possible for Ptolemy to keep to Aristotle's basic belief that the earth is at the centre of the universe (Figure 17). Ptolemy's picture of the universe dominated astronomy for 1400 years. No-one questioned it seriously until the time of the Renaissance.

5 Copernicus (Nicolaus Kopernik 1473–1543)

In 1543 a revolutionary new theory was published by Copernicus (Figure 18). He was a Polish scientist and also an official in the church.

He believed that our common-sense idea that the sun rises and sets could be explained in a different way. He believed that the sun did not move round

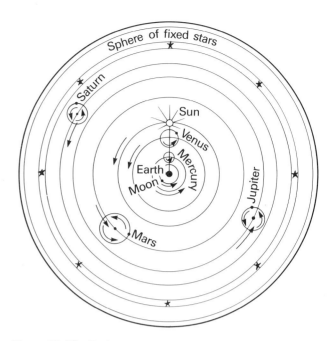

Figure 17 The Ptolemaic universe

Figure 18(a) Copernicus, who was a canon in the Roman Catholic Church, called Ptolemy's picture of the universe monstrous

Figure 18(b) This map shows Torun, Copernicus' birthplace, and Frauenburg where Copernicus lived for much of his life

the earth but that the earth moved round the sun. Because the earth turns round, one side faces the sun during the day while the other side remains dark, and then vice versa.

Bringing in the Copernican revolution

Copernicus' sun-centred theory of the universe
Figure 19 is a modern drawing of the Copernican system. (Compare this with the Ptolemaic universe in Figure 17.)

1   The sun has been placed at the centre of the universe and the earth has become one of the planets which orbit the sun.

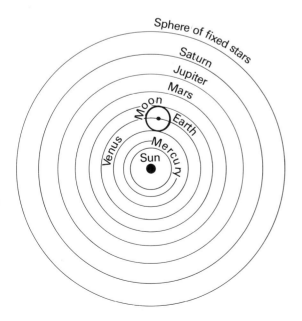

Figure 19 The Copernican universe

2   Copernicus did not completely break with tradition. Like Aristotle, he still thought in terms of spheres, with the sphere of the stars outside the spheres of the planets.
3   Only six of the nine planets were known at that time. Which three planets are missing?

Copernicus suggested that the earth moves in two main ways.

1   The earth spins on its axis (Figure 20). This accounts for the movement of the sun and stars across the sky, day and night. It spins once every 24 hours.

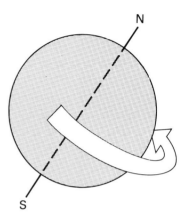

Figure 20

2   The earth orbits the sun, each orbit taking 1 year (Figure 21). This accounts for the way in which we see the movements of the planets from earth – the problem of the 'wandering planets'.

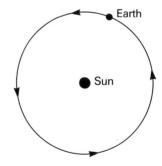

Figure 21

How Copernicus arrived at his theory
Copernicus wrote down his new theory in a book called *On the Revolutions of the Heavenly Spheres.* He dedicated his book to Pope Paul III, and in his introduction he describes how he hit upon his new idea of a sun-centred universe. This is what he said (Figure 22):

'How I came to dare to think that the earth *moves*
and so contradicts the received opinion of the
mathematicians and indeed to contradict the
impression of what we see with our eyes, is what
your Holiness will want to hear. So I should like
your Holiness to know that it was this. I realised that
the mathematicians are reaching contradictory
conclusions for they have either omitted some vital
detail or introduced something foreign and wholly
unnecessary . . . .'

'So I pondered long upon this uncertainty of
mathematical tradition in establishing the
movements of the system of the spheres . . . .'

'I therefore took pains to read again the works of
all the philosophers that I could lay hands on to seek
out whether any of them had ever supposed that the
movements of the spheres were other than those
demanded by the mathematicians. I found . . . that
Hicetas had realised that the earth moved.
Afterwards I found in Plutarch that certain others
had held the same opinion . . . .'

'Taking advantage of this I too began to think of
the movement of the earth; and . . . knowing now
that others before me had been granted freedom to
imagine such circles as they chose to explain the
phenomena of the stars, I considered that I also
might easily be allowed to try too. If I assumed that
the earth moves I might be able to give a sounder
explanation than the mathematicians give and so
explain better the movements of the celestial
spheres.'

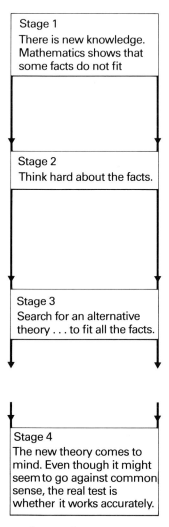

Figure 22 Stages of a new theory

6 Galileo Galilei (1564–1642): 'A new world through a telescope'

At first, very few astronomers took any notice of
Copernicus' new theory. It was more than 50 years
after he died that the main breakthrough took place.

In Padua (Italy) in 1609, Galileo (Figure 23) was
starting out on an important area of discovery. He
had heard about the invention of what we now call
the telescope and had realised at once what a useful
instrument for astronomy that would be. He
was already convinced that Copernicus' theory was right,
but what he saw through his telescope made him
even more certain.

Aristotle had said that the heavens are perfect but,
when Galileo turned the telescope on the moon, he
could see that it 'is not robed in a smooth and

Figure 23 Galileo Galilei

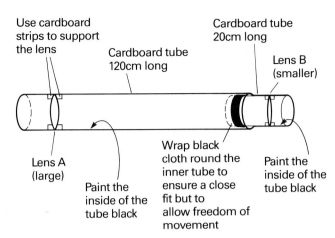

Use cardboard strips to support the lens

Cardboard tube 120cm long

Cardboard tube 20cm long

Lens B (smaller)

Lens A (large)

Paint the inside of the tube black

Wrap black cloth round the inner tube to ensure a close fit but to allow freedom of movement

Paint the inside of the tube black

Figure 24 Making a simple telescope like that of Galileo

polished surface but is in fact rough and uneven, covered everywhere just like the earth's surface, with huge mountains, deep valleys, and chasms'.

Even more surprisingly, he discovered that the sun was far from perfect and that in fact it had spots! Through his telescope, Galileo discovered that the planet Jupiter has its own moons. It appeared to Galileo just like a miniature model of Copernicus' solar system. Here at least was proof that *not everything* revolves around the earth, because Jupiter's moons revolve around Jupiter.

In addition to his work with the telescope, Galileo studied mechanics – the theory of how objects move. He soon realised the importance of mathematics and mechanics in helping us to understand the workings of the earth and also the workings of objects in the heavens.

Putting all these discoveries together, Galileo became convinced that Copernicus' theory was sound, but Galileo went one step further. He said, 'This theory explains the "appearances", but it is not just a theory, it is how things actually are: it is a proved fact.' He placed these new-found 'facts' in the context of Aristotle's theory in order to correct Aristotle.

There was much opposition to Galileo's picture of the world. Many leading scholars continued to think that Aristotle was right and that Galileo had not *proved* his point. In the end, Galileo was arrested.

On 12 April 1633, Galileo was tried before the Church Authority in Rome. He was an old man; his eyesight was failing, and he was ill and rather frail.

Galileo was charged with disobeying an order not to teach or defend Copernicus' theory of a sun-centred universe. He was dismayed! He was a loyal

member of the Roman Catholic Church and was sure that the new astronomy was not a threat to the Christian religion. For more than 20 years he had been observing the skies with his telescope and everything he saw convinced him that the earth is a planet which orbits the sun. Why, he asked himself, are the leaders of the church so blind? Some would not even look through his telescope, while others blamed the crude telescope for distorting reality.

Galileo was sentenced to imprisonment. He spent the rest of his life as a prisoner in his house in Florence. While he was there he wrote his most important scientific book on the theory of how objects move.

Galileo's book compared the earth-centred system of Aristotle and Ptolemy with the sun-centred system of Copernicus. To prove that Copernicus was correct, Galileo used his knowledge of physics and his actual observations through his telescope. 'You can see that it moves,' he said.

Galileo was condemned by the theologians in Rome for teaching (contrary to the Bible) that the earth moves. He was also condemned by the University of Oxford and many scholars.

They did not mind so much the fact that people proposed theories about the stars and the planets, but they did mind when Galileo said that 'the earth actually moves'.

Galileo died in 1642, the year that Isaac Newton, one of the greatest scientists in history, was born.

7 Johannes Kepler (1571–1630)

In the same year as Galileo first used a telescope to see the stars, another important discovery was made. Johannes Kepler knew that Copernicus' theory was not completely accurate. He concentrated mainly on the orbit of Mars. After 10 years of work on the problem, he suddenly hit on the idea that the orbits of the planets are not perfectly circular, but elliptical. (An ellipse looks like a circle that has been slightly flattened.)

Kepler was lucky in having many accurate observations of the planets that had been made by a brilliant Danish astronomer, Tycho Brahe. When Kepler checked his idea of elliptical orbits against Tycho's observations, he found that it fitted them beautifully. This was a great breakthrough, for it removed one of the most damaging criticisms of Copernicus – that his theory could not be made to

Science: changing our view of the world

Scientific discoveries are made in many different ways, but often the pattern of discovery is similar to that described by Copernicus (Figure 25). After studying long and hard over a problem, the new idea springs to mind in a moment of creative insight. Can you think of any occasions when *you* have been struggling with a problem, perhaps trying to understand something that you have been learning at school? You cannot see your way through it and then, quite suddenly 'the penny drops', as we say.

The same is often true in religion.

(a) A breakthrough in technology – the discovery of fire

(b) Copernicus discovered a secret of the universe – the fact that the earth is only a planet

(c) Darwin introduced the concept of evolution to explain diversity in living things

Figure 25 Revolution in science

(d) Einstein discovered a hitherto unknown fourth dimension

Religion: Changing our view of the world

Every now and then in the history of both science and religion someone has suddenly come to see the world or some part of it in a different way. In religion, such changes of view are concerned with understanding of the world and the meaning and purpose of life (Figure 26). In science, they are to do with the structure of the world and how it works. In both science and religion, however, such changes usually stem from someone with vision – someone who see things in a new way and not the way that things have been understood in the past. The new view or picture of the world may be so powerful that it changes the course of history.

The pictures in Figure 26 show important moments of revelation from four of the world's main religions. See whether you can find out what happened in each case and why they were so important in the history of religion.

(a) Santa Katarina monastery on Mount Sinai where Moses received the Ten Commandments

(b) The conversion of Saint Paul

(c) Gautama Buddha who founded Buddhism is said to have had a revelation under the Bodhi tree or tree of knowledge

(d) In the centre of Mecca is the Ka'aba – the sacred black stone that all Muslim pilgrims must kiss

Figure 26 Moments of revelation in religion

work accurately. The heavens, it seems, work according to mathematical laws. Kepler was able to summarise his discovery in three basic laws of planetary motion. In 1609 he published his discovery of elliptical orbits in a book called *On the Motion of Mars*.

8 Isaac Newton (1642–1727)

In the history of science there have been a small number of scientists whose work has been especially important. One of these was Sir Isaac Newton. Even the greatest scientists, however, realise that they owe much of their success to others who have gone before them. Newton completed the work of Galileo and Kepler and also secured the victory for Copernicus against Aristotle and Ptolemy.

Figure 27 Newton's home in Lincolnshire where he did some of his most important work

One day when Newton was at his home in Lincolnshire, he saw an apple drop to the ground. In a moment of inspiration he realised that the 'force' which pulled the apple to the ground is the same 'force' which keeps the moon in its orbit so that it does *not* fly off into space. He made a leap of imagination to join into one force two events which seem to be the opposite of each other, and to use mathematics to measure that force.

Kepler had discovered that planets move in elliptical orbits around the sun according to strict mathematical laws, but he could not explain why. Galileo's study of mechanics had shown that, if an object is made to move, it will continue to travel at the same speed and in a straight line unless it is acted on by a force. Newton joined these two ideas together in his theory of universal gravitation.

The paths taken by a projectile launched at various speeds from a great height

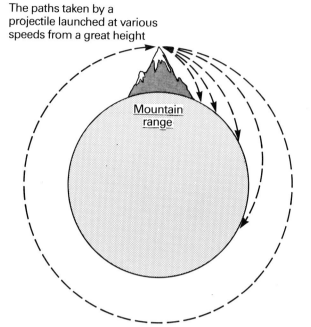

Figure 28 The effect of gravity on a projectile

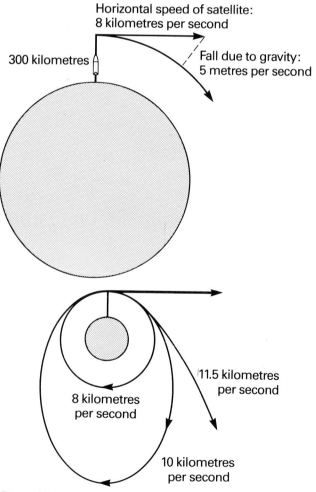

Figure 29 Newton's diagram showing what would happen to a projectile launched from a great height at different speeds

Newton imagined what would happen if an object were fired from a great height parallel to the earth's surface at different speeds (Figure 28). At slow speeds the earth's gravity would cause the object to fall to the earth but, if the speed were great enough, it would continue right round the earth to where it had been fired. At even greater speeds, it would leave the earth completely.

Newton's law of universal gravitation is used every time that a satellite is put into orbit around the earth. To launch a satellite to, say, 300 kilometres above the earth, you could first fire it horizontally at a speed of 8 kilometres per *second*. In the first second it would travel 8 kilometres horizontally but it would also fall 5 metres because of gravity. However, the satellite would be no closer to the earth because the earth's surface also falls away from the horizontal by 5 metres every 8 kilometres. The satellite would be in orbit parallel to the earth's surface.

At 8 kilometres per second, the orbit is circular.
At 10 kilometres per second, it will be eliptical.
At 11.5 kilometres per second, the satellite will escape the earth's gravitational pull completely.

9 Religion and ways of knowing

Turn back to p.21, and copy and complete Table 1. Then look again at the four different ways of knowing. We have seen that these are all used in science. They are also used in religion.

The subject matter of religion (which is to do with the purpose of life and belief in God) is even more complex than that of science. It is therefore even more difficult to prove what is right. Yet this does not mean that no-one *can* be right. For example, you can be correct in thinking that someone loves you even though you cannot *prove* it to anyone else. Make a list of situations in which you can say 'I know, even though I cannot prove it to you'.

Today, many people do not take religion seriously because it is difficult to prove its truth. They tend to say, 'No-one can know – it's all a matter of opinion'. But how do they know that no-one *can* know? In the past, many people said this about natural events which scientists *can* now tell us about.

Religious people often claim to have great certainty. How do we *know* that they are mistaken?

If a scientist makes a claim about the physical world, then we investigate the claim by examining the scientific evidence. Similarly, we can investigate a religious claim by examining the religious evidence.

Just as science has progressed through the insights of great thinkers, so has religion. Some people have very deep religious experiences, and develop great insight. What such people have to say about religion can be taken as evidence. The lives of such people can also be weighed up as evidence in deciding whether religion contains truth or not.

3
A terrible storm

A day on the beach

*T*HE camp site was some 3 miles from the nearest beach, and everybody was hoping that the weather would be warm and sunny. The next day was gorgeous, and so they went by foot to the sea and spent the entire day there.

A number of the campers ventured into the water. Others not so keen on swimming, decided to go for a walk along the base of the cliffs, at the top of the beach. They had been gone some time when they returned carrying some fossilised shells which they had dug out of the soft rock.

'What are these, Miss?' enquired one of the walkers, as they approached the teachers.

'What do you think they are?' replied Miss Ridgewell.

'Well, they're fossils, but what kind?'

She took some of the samples and studied them for a while.

'These are ammonites. How old do you think they are?' she asked.

(a) A fossilised ammonite

The rock pools contained a seemingly endless array of life.

(b) The ammonite as it would have looked 180 million years ago

'I don't know. About a million years I suppose,' said one boy, expecting that he had probably overestimated the age.

'About 180 million years would be my guess,' she replied.

The boy looked astonished.

'I can't even think that long,' he said.

By late afternoon, dark clouds began to appear and the air became motionless and humid, but nobody took much notice. When the other visitors had left the beach, there was plenty of room for a game of cricket. As the game ended they noticed that the tide had gone down sufficiently to reveal rocks and a host of small pools, left behind by the receding water. They all went to investigate. At first it seemed that there was little to see in the pools but, after longer and closer inspection, they began to notice an endless array of life. It was helpful having Miss Ridgewell with them, because she was able to point to things of interest which the others might have missed.

Their fascination with the rock pools prevented them from noticing how the skies were darkening quite rapidly – not so much because of the approach of night, but more because of the thickening cloud.

'We had better be getting back,' said Mr Payne. 'It looks like rain and it will take us a while to return to the camp.'

'It's creepy'

They gathered their things together quickly and set off back along the path. Their route took them across three or four grassy fields, then along a country lane for about a mile and finally through the middle of a dense wood. By the time that they reached the edge of the wood the light was fading fast, and under the covering of the trees it became almost as dark as night.

'It's creepy in here,' said Dawn nervously. 'The trees look like giant animals.'

'Don't be daft,' taunted Wayne. 'There's nothing to be scared of.'

'You can almost imagine faces looking at you through the trees,' observed Carol.

'Don't say that,' said Dawn. 'I'm scared enough as it is.'

The walk back through the woods seemed endless, and some of them began to wonder whether they were on the wrong path. So they all felt a sense of relief when they finally broke through and were actually back in the camp site.

A storm

Realising that a storm was on the way, Mr Jones asked them all to gather in the stores tent with their sleeping bags.

'I suggest that you find yourself a space and we'll see whether the storm passes by,' said Mr Payne.

They had barely settled before the rain started to fall more insistently; the storm was indeed coming their way. Inside the tent there was noisy chatter – a mixture of excitement, tinged with the feeling that they were not as well protected as they would have been at home.

As the storm worsened, lightning tore through the night sky, seeking out some unsuspecting object on earth, and cracks of thunder whiplashed the air. With each brilliant flash of light the interior of the tent was momentarily illuminated as bright as day.

Their feelings of uncertainty and insecurity increased. It no longer seemed quite the fun that it had at first. Most of the youngsters hid in their sleeping bags or covered their faces with their hands. Some screamed out at each crash of thunder, but a few could not resist looking through the gaps at the front of the tent. At the height of the storm, when the rain was falling in torrents and the wind threatened to carry the canvas tent across the field, there was one tremendous flash of lightning, followed instantaneously by a deafening clap of thunder. For a moment they were all terrified. 'That must have been pretty close,' said Mr Payne, quietly, his voice betraying concern for their safety. Nobody felt like answering. By now they just wished the storm would go away.

'I think that must have hit something close to us, perhaps one of the trees,' said Mr Jones to the other teachers. 'I'll be glad when it's over.'

The storm raged for a full hour before the rain began to ease and the thunder rolled on to somewhere in the distance. They could still hear it, and the lightning still brightened the sky but, for those huddled beneath the canopy of the tent, the terror had gone. The tension eased, and gradually the talking began again.

'I've never been in a storm like that before,' said Winston. 'You notice it worse out here, don't you?'

The problem of natural disasters

'What I can't understand', said Carol, 'is this. If the world is meant to be so wonderful and created by God, why are there things like lightning and earthquakes? They don't do any good. They just kill innocent people, don't they?'

'I know what you mean,' replied Mr Payne. 'Some people find it hard to believe in a creator because of these kinds of natural disasters.'

'Strangely enough,' replied Miss Ridgewell, 'that's how many scientists believe life on earth was created. Earthquakes, for example, have played an important part in the formation of th earth's surface.'

'So what you're saying is that, if there hadn't been things such as earthquakes and lightning, we wouldn't have been here?' said Winston.

'That still doesn't answer my question, though, does it?' said Carol. 'Surely, if there is a God, he could have created the earth without using these ways of doing it.'

'I don't know,' replied Miss Ridgewell.

'That's the mystery of it, isn't it? Perhaps everything could have been different or perhaps this is the only way that it could have been created. I can't answer that question. What do you think?'

The creation of life on earth

One most widely held theory suggests that in the early stages of the earth's history the atmosphere was thin and consisted of hydrogen, carbon monoxide, methane and ammonia but little if any oxygen. Ultra-violet light from the sun penetrated to the earth's surface and volcanoes spewed ash and lava into the air. Water vapour had condensed to form the seas. Life was formed when this early 'soup' was bombarded by fierce electrical storms.

Things to do

Thinking about the story

1 Have any of you ever experienced something which really frightened you? Can you describe how you felt at the time?

2 If God created the world, do you think that He could have done it differently? Can you think of any changes that *you* would have made if you had created the world? Would your changes lead to any other problems?

Creative responses

1 Make a sketch or paint a picture of the storm as you imagined it in the story.

2 Write a piece of poetry to describe the picture of the storm which you see in your mind.

3 You could also use this story of a storm to do some drama in class. In your acting try to re-create the feeling of tenseness and excitement as the storm rages.

Topic work: The earth beneath our feet

1 Fossils and the story of Mary Anning

The story of Mary Anning (Figure 1) begins in the coastal town of Lyme Regis. Her father was a cabinet maker who collected and sold fossils to make a little extra money to support his family. In 1799 his wife Mary had a baby daughter who they also called Mary. When she was only 15 months old, she had a very lucky escape. A nurse who was carrying her was struck by lightning and killed, but Mary, although left unconscious, recovered.

Figure 1 Mary Anning

As Mary grew up, she enjoyed walking along the beach and the cliffs, helping her father to find fossils but, when she was only 11 years old, her father died. Mary and her elder brother Joseph continued their fossil hunting, and a year later Joseph discovered what he thought was the fossilised head of a crocodile embedded in the limestone cliffs. It had a huge jaw and menacing teeth. More of this creature became visible after a violent storm washed other parts of the limestone away. Mary, who was still only 13 years old, took a hammer and chisel and with great skill traced the outline of the fossil. They could barely believe their eyes – it was unlike anything that they had ever seen before (although, in fact, other specimens had been found). The strange monster,

Figure 2 An ichthyosaurus

whose remains had been fossilised in the rocks some 200 million years previously was later identified as an Ichthyosaurus (Figure 2). It was an ancient reptile which lived in the sea, feeding mainly on fish and cuttlefish. At the time, nobody knew how long it had been there.

Not satisfied with this exciting find, Mary went on to become one of the most famous collectors of fossils in the early nineteenth century. Some of the most eminent geologists accompanied her on her expeditions. She seemed to have a special talent for knowing where fossils were likely to be found. In 1824 she discovered the first complete plesiosaurus and then she found another in 1830.

Her most remarkable find was made in 1828 when she uncovered the remains of an early form of Pterodactyl which was named a Dimorphodon.

Mary Anning died of cancer when she was 48 years old but, although her life was short, she had made an important contribution to the development of geology. On her death, the Geological Society (which had been founded in 1807 to advance the study of geology) paid her the unique honour of recognising her work, even though she was not a member.

Nowadays everybody has heard of dinosaurs and we tend to take the fact that they once inhabited the earth for granted. Think about the story of Mary Anning, and try to imagine how she, and others of her time, felt when the remains of these animals started to be found.

- What might she have thought that these animals were?
- How might she have explained the fact that these animals no longer exist?

2 More on fossils

Fossils are the remains or traces of animals and plants which lived long ago. Most plants and animals die, decompose and leave no trace but, sometimes when a plant or animal decays in mud, the tougher parts such as twigs, or bones or shells survive in exactly their original shape. So we get leaf imprints in rocks (e.g. in coal). Very occasionally, the whole skeleton of an animal is preserved untouched for centuries in this way.

Vast collections of a variety of fossils have been dated in different layers of rock. By putting these in order – from the very oldest fossils to the more recent ones – scientists have been able to find out about the shape, size and development of different species of plants and animals. In other words, rocks provide a historical record of successive ages of the earth.

Scientists have compared in detail an enormous number of fossils all over the earth. So the 'fossil record' of the rocks has made it possible to date the story of living creatures (see Table 1).

The fossils record shows a similar succession for plants.

Table 1 Fossils of animals

Date	Fossils of animals
3 000 000 000 (3000 million) years ago	Single-cell animals (e.g. bacteria)
	Multiple-cell animals without backbones (e.g. jellyfish)
	Fish with backbones; amphibians (creatures which live in water and also on land); reptiles (including dinosaurs)
100 000 000 (100 million) years ago	Birds and mammals

> **Geology** is the study of the earth's crust, especially of the rocks forming its surface.

The dating of rocks

The dating of fossils has in turn been used by scientists to date different layers of rock and to make geological maps.

The first geological map in the world was drawn up in 1815 by William Smith, a canal engineer working on the Somerset Coal Canal. He noticed that the *different* layers of rock contained their *own* kinds of fossils. Later, when he had worked on canals in other areas, he realised that rocks which have the *same* kinds of fossils are of the *same* age, even though they are in different parts of the country. This was an important breakthrough in the study of geology and provided a method of dating the age of different layers of rock. The layers are called **strata** and this method is called **stratification**.

The types of fossils which William Smith found were mostly the remains of simple sea animals. A similar method of dating rocks by the fossils which they contain was being used in France by Georges Cuvier (1769–1832). Cuvier concentrated on the fossils of land animals because he was interested in using the remains to show what the ancient animals had looked like before they became extinct. Smith used his method to map out the different bands of rock across the whole of England (Figure 4).

William Smith was a Christian. He saw a close connection between his geology and his religious beliefs. The sheer beauty of the fossils and the order of what he was discovering made a great impression on him. It strengthened his belief that God had created it all.

He did not take the Bible stories literally. The creation, he believed, was not completed in 6 days of 24 hours each but by gradual processes. He thought that each day in the Genesis story represented a long geological age. The earth's surface had changed and developed, but for Smith this was still the result of God's creative action through the natural laws.

How continents are formed

Fossils were also used as evidence for the modern theory of continental drift which has revolutionised the study of geology. **Continental drift** is the name given to the way in which we now believe continents have been formed.

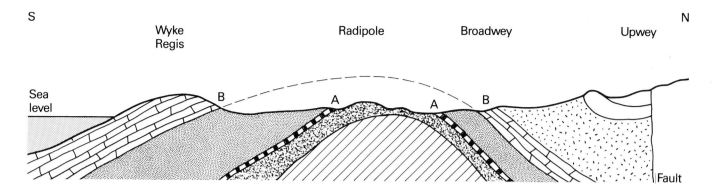

Section across the Weymouth anticline, Dorset,
showing Jurassic rocks 140 million years old

Figure 3 Using fossils to match rock types

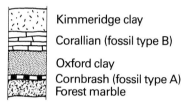

Kimmeridge clay
Corallian (fossil type B)
Oxford clay
Cornbrash (fossil type A)
Forest marble

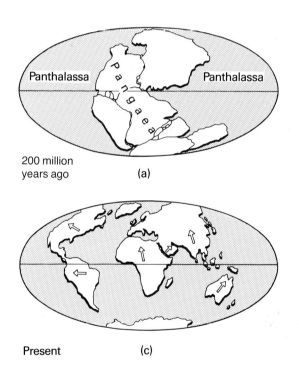

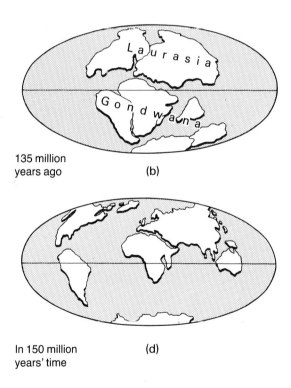

Figure 4 Maps showing the break-up of Pangaea

The person who put this theory forward first was Alfred Wegener (1880–1930) exactly 100 years after Smith's geological map. At the time, other scientists could not accept it because it seemed ridiculous, yet it was to be a most important discovery. (Does this remind you of Copernicus?)

Wegener was not the first to notice how the continents of North and South America fit alongside those of Africa and Europe. A number of theories had been suggested to explain this 'fit'. In 1915 Wegener put forward the idea that they had once been joined to each other and had gradually drifted apart. He gathered together many facts to support his theory. He pointed especially to the way in which fossils found in Brazil were similar to fossils found in the 'matching' part of Africa. Why might this suggest that the continents had once been part of the same land mass?

This kind of discovery about the world is very satisfying. It is rather like doing a jigsaw. It is fun to do. It is also enjoyable to see the finished picture *and* the way that all the separate pieces fit together.

Some geologists were interested in Wegener's theory to begin with but it was such a revolutionary idea that one of his opponents said, 'If we are to believe Wegener's hypothesis, we must forget everything which has been learned in the last 70 years and start all over again.' While the majority of geologists refused to believe Wegener, a few were convinced and managed to keep his ideas alive. Only in the 1960s were Wegener's ideas finally taken seriously, many years after he had died.

How earthquakes happen

The theory of continental drift *has* revolutionised the study of geology. With this theory we can now explain how the earth's surface is gradually changing shape and why events such as earthquakes only happen at certain places on the earth's surface. They occur where the sections of the earth's crust carrying the continents collide and rub against each other (Figures 5 and 6).

3 Natural disasters and religious belief

On the morning of 1 November 1755 the churches in Lisbon, the capital city of Portugal, were full of people. It was All Saints Day – an important Christian religious festival. At 9.30 am a great earthquake occurred, one of the largest ever recorded. Suddenly, and without warning, the buildings began to shake and then came crashing to the ground (Figure 7). Thousands of people were killed, particularly those in church who were crushed by falling masonry. The devastation was made worse by a fire which broke out shortly afterwards. It gutted much of the city. The earthquake was so violent that it was even felt in England.

Figure 5 The sections or 'plates' which make up the earth's surface, and the directions in which they move

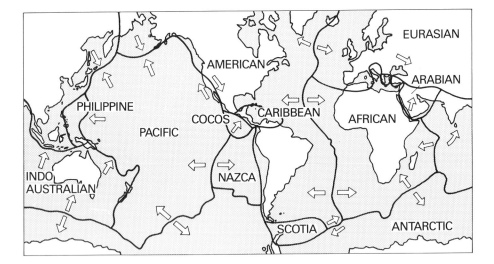

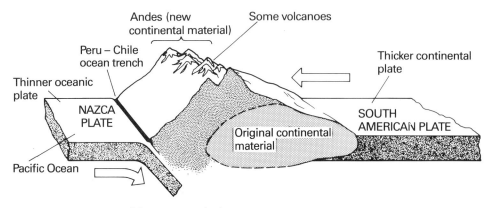

When an oceanic plate collides with a thicker continental plate, it sinks beneath the continental plate. This is called subduction. Here the Nazca plate sinks beneath the South American plate. Sediment from the ocean floor is deposited on to the edge of the continental plate to form the Andes. The friction which occurs as the two plates slide over each other may cause earthquakes and volcanoes.

Figure 6 Subduction of the Peruvian Andes

Figure 7 The Lisbon earthquake

Europe was stunned by the disaster. How could such a thing happen? Was it a punishment from God? The fact that so many of the victims had been in church caused particular problems in trying to come to terms with the disaster. 'How could a loving God allow this to happen?' they asked.

- How do you feel about natural disasters like this?
- Do you think that they are strong arguments against the idea that there is a loving God?
- Does it make any difference to your answer to know (as we now do) that earthquakes are the result of the natural processes which have formed the earth?
- If there is a God, *should* he have formed the earth so that these kinds of things did not happen? *Could* he have formed it without using these kinds of natural processes?
- Can you think of any way in which we could avoid the worst effects of these 'natural' events?

In Alaska in 1964 on Good Friday, another violent earthquake occurred (Figure 8) which was almost as great as the one at Lisbon. Here is one woman's eye-witness account of the disaster.

'It was a little after 5.30 am that I heard a rumble. I had heard one before, just preceding a mild earthquake last summer, but we also hear frequent rumbles from the big guns firing at the Army base.

Something instantly told me that this was another earthquake. I leaped off the bed, yelling 'Earthquake!' I grabbed Anne and called to David. They both moved with lightning speed. We had reached the front hall when the house began to shake.

Within a few seconds the entire house started to fall apart, splitting first in the hallway that we had just come through. We heard the crashing of glass and then that horrible rending sound of wood being broken apart. The trees were crashing all about us, adding to the terrible din.

I looked towards the car to see whether it was shaking as much as during the last quake and, as I watched, the garage collapsed on top of it. Now the earth began breaking up and buckling all about us. A great crack started to open in the snow between Anne and me, and I quickly pulled her across it towards me.

Then our whole lawn broke up into chunks of dirt, rock, snow and ice. We were left on a wildly bucking slab; suddenly it tilted sharply, and we had

Figure 8   The Alaskan earthquake

to hang on to keep from slipping into a yawning chasm. I held David but Anne had the strength and presence of mind to hang on by herself. Although crying, she was still able to obey commands – thank God, because poor Dave was hysterical, and I could only hold him tightly.

By now, both the children were hysterical, crying and saying over and over, "What will we do? We'll die . . . ." I knew that we'd have to move now, carefully, but fast. I had to find a way up that cliff, and we should have to climb over the great chunks of earth without falling into holes and crevasses.

I suggested that first we say a prayer asking Jesus to take care of us and guide us, and both children stopped crying, closed their eyes and fervently pleaded with Him to take care of them. This had an extraordinary effect on them and on me. Anne was ready now to climb on her own and, although David was still worrying about his bare feet and frostbite, he had stopped crying.

A man appeared above the cliff. All three of us immediately yelled, "Help, help – come and get us." He shouted down that he would find some rope and then disappeared. Suddenly six or eight men appeared at the top of the cliff. One man, whom I still have not identified (a great pity because we feel eternally grateful to him), started down the cliff towards us. The children both hugged our rescuer, and I could feel their relief as they told him how cold they were. He put his black wool jacket around Anne

and for a week she wore it almost constantly. It is dirty and worn and much too big, but it will be her most prized possession for a long time to come.

The strongest feeling of all I know I share with the children and thousands of other fellow Alaskans: a fervent thankfulness to God for having spared our lives in one of the world's worst earthquakes. We are thankful for the opportunity to rededicate our lives to His service.'

This story shows clearly how frightening it is to be caught up in these kinds of disaster.

- Having read the story, can you list what they were most frightened about?
- What effect did the woman's religious beliefs have on the situation?
- Did she blame God?
- Can you understand why she took the view that she did?
- What did she mean by 'rededicate our lives' in the last paragraph?
- How do you think that she would have responded to Carol's question (on page 37)?
- What do you think about the incident? Was her rescue just coincidence? Could the woman have been mistaken in thinking that God spared their lives? If she was not, what does this say about God, and how God regards natural disasters?

4 For discussion: possible 'answers' to the problem of suffering

On page 43 you were asked some difficult questions. This section puts forward various ideas and things to think about. When you have studied it, go back to the answers you gave to those questions. Do you still think the same about them? Try to explain your reasons.

It is most important to appreciate that deep questions such as these cannot be answered once and for all. The greatest thinkers who have ever lived have gone on wrestling with possible answers. It is important, however, that we think out our own answers to them and learn too from other people's experiences and the way that they see situations.

Laws of science

Things in this world are limited, that is to say, they have particular properties, and not others; everything is one thing and not something else. A piece of paper does not have just one side; a horse does not suddenly become as light as a feather when it treads on a daisy (Figure 9); if a man falls off a mountain ridge, he does not sprout wings like a bird; the apple which fell on Isaac Newton as he sat under the apple tree did not become soft when it came into contact with his head.

Science would be impossible in a world in which such things did happen. So, far from being unpredictable, the way in which things work occasionally results in tragedies. If one thing happens to meet another thing in particular

Figure 10 Two cars arriving at the same place at the same time

circumstances (e.g. two cars happening to arrive at exactly the same place at the same time), a crash must occur.

This allows for the ordinary 'factual' explanation of how things happen. A storm (Figure 11) or an earthquake (Figure 12) can be understood in these terms.

Figure 9 The horse and the daisy

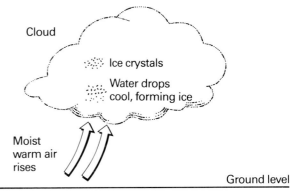

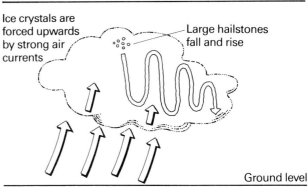

As the hailstones vibrate against each other, they produce static electricity.
Large amounts of static electricity produce lightning.

Figure 11 The formation of a storm

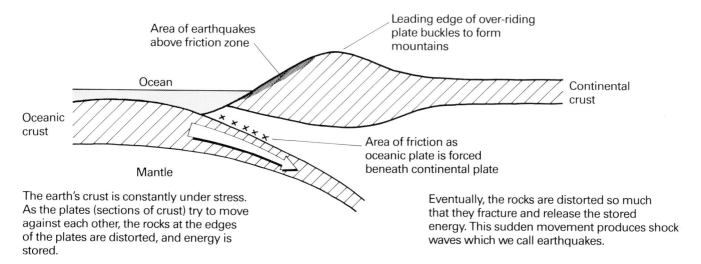

The earth's crust is constantly under stress. As the plates (sections of crust) try to move against each other, the rocks at the edges of the plates are distorted, and energy is stored.

Eventually, the rocks are distorted so much that they fracture and release the stored energy. This sudden movement produces shock waves which we call earthquakes.

Figure 12 The formation of earthquakes

A paradox: suffering as both bad and good
The fact of the 'laws' of science does not answer Carol's question as to *why* the world was made in this way. What we call the problem of suffering is to do with the harm and the pain that the world causes people, and animals too. This brings in a different set of considerations.

It is interesting that, when people think about suffering, they often see two things about it which seem to contradict each other. They see that suffering is bad and should be avoided if possible, but they also see it as helpful, in a strange way, in that it develops courage and patience and in this sense it is good. This is a **paradox**, a term used when two statements are felt to be true and yet contradictory.

Thus suffering is seen as a paradox by many people. No-one doubts that it is bad.. The damage which a storm or earthquake can do is horrific. It cannot be good that people are killed or paralysed. Similarly, it cannot be good that people are ill. Yet suffering can also be seen as good for people.

Here are some other examples of paradoxes from well-known sayings.

Paradox A   'Make haste slowly.'
Paradox B   'Many hands make light work.' 'Too many cooks spoil the broth.'
Paradox C   'The most incomprehensible thing about the universe is that it is comprehensible' (Einstein).

- Explain how each of these ideas seems not to make sense and yet may be true.
- Think of other examples of paradoxes.

Study the following paragraphs (a)–(d) carefully and see whether you agree or not, and why.

(a) The need for challenge and excitement
Danger, real danger, is exciting. It makes people alert and fully alive. It trains character. It teaches people to discipline themselves and to co-operate with others. In fact, people often seek danger. So they go potholing, skiing, mountaineering, sailing and so

Figure 13 People often seek danger

Figure 14 Leprosy destroys nerve ends and therefore feeling

forth. It appears that people need something to fight against in order to develop properly as people. It is important for us to note that there would be no danger or challenge if the possibility of suffering were not real.

(b) Nervous system You may like to consider that suffering is the result of the fact that we have a nerve system. This is essential if we are to avoid damage to our tissues. Leprosy kills the nerves and so lepers suffer severe damage to their tissues because they cannot feel when they cut themselves. It seems essential for our well being that we have nerves and yet, paradoxically, they cause us to suffer.

(c) Learning to help and not to hurt other people
There is a deeper level at which most people would agree that suffering, while bad, can also be good. It is a means by which people can learn to respect, help and love each other. People are often very selfish and caught up in themselves and hardly notice anyone or anything else. But if suffering comes to them, it can give them more sympathy for other people; they can become more aware of how difficult life can be for everyone. Also, if people suffer, it gives others an opportunity to be kind and helpful. Because people can cause suffering in others, it gives them the responsibility not to hurt, but to help. We are free to choose to be nasty or nice.

(d) Developing a strong character Besides helping people to learn to co-operate and to have good relationships with each other, suffering can help a

person to rise above himself or herself. It can develop courage, patience, determination and resourcefulness – real strength of character. Some of those people who have suffered most have become the greatest people.

‘Heaven’ is the purpose of life
For religious people, this world of time and space is like a school to prepare people for eternal life. People have bodies which can feel pain so that they may learn such qualities as patience and endurance. **Heaven** is the name often given to that eternal and completely satisfying pain-free life.

Figure 15 Bob Geldof was the inspiration behind Band Aid which raised millions of pounds for people in Ethiopia

Is 'heaven' a place?
When the Buddha was asked where heaven is, north, south, east or west, his answer was a question: 'when a fire goes out does it go north, south, east or west?

Many people believe that a human being is merely the body which they see and the brain on which a surgeon could operate – this is what can be explored by science. Others believe that people are also spiritual, and they find that this helps to make sense of the nasty side of life as well as all that is nice. Thus the terror felt by Dawn in the wood, the danger of the storm and the horror of the earthquake can be a means of developing the spiritual side of a person.

Mystery in science and religion

A 'mystery' may be like a detective story which has to be solved, or it may be something which awakens a sense of wonder. The cleverest scientist knows that he or she is still extremely ignorant; our small brains and short experience can enable us to grasp only a few things.

Einstein, perhaps the greatest scientist of the twentieth century, used to quote a parable of Isaac Newton: he felt that he was like a little boy playing beside the vast ocean of truth and picking up a few pebbles. People who are 'know-alls' are generally very small-minded indeed and very ignorant because they do not know how much they do not know. People who are really clever know how little they know and are always eager to learn more.

A scientist can, for example, use a telescope or a microscope to explore *how* the universe works, but there are some questions which the telescope and the microscope cannot answer. Yet these are the questions which have exercised the minds of the greatest people who have lived, and the questions which almost everyone goes on asking. Because a mystery is something we can ask more and more about and never finish learning about, it makes people determined to go on seeking answers to such questions.

This section has given some possible 'answers' to one such question, i.e. the problem of suffering. Many people do not think these 'answers' are helpful. The problem of suffering still prevents some people from believing in God, and yet there are perhaps even greater problems if God is ruled out.

- Why is there such beauty and order in the world?
- Why do we enjoy finding out about the world so much?
- Why do we value kindness and love and courage and determination and so many other good qualities in people?

Think about all these ideas for yourself and at different times. A true answer must be one which stands up to all the facts that we experience.

4

A day on the farm

A lucky escape

*T*HE morning after the thunderstorm the air was decidedly chilly and damp. There was little enthusiasm to get up on such a dreary morning. But all that changed when one of the girls called out from across the field.

'Look at this!' she shouted. 'That's what we heard last night in the storm. It's a tree. It's been struck by lightning.'

The response was immediate. Curiosity overcame their tiredness and there was a stampede out of the tent. A large branch of the tree had been ripped from the trunk.

'We weren't far from that, were we?' said Leila. 'It could have hit our tent, couldn't it?'

'It just shows you how important it is to keep well away from trees when you're camping,' said Mr Jones with an air of authority in his voice. 'Anyway, we are all up now; so let's get on. I don't want to be late getting to the farm.'

Working with animals

When they had finished their breakfast, they set out through the wood and along the edge of a large field. The mist had cleared and the sun had broken through the clouds. As they neared the farm, they could see that the sheep had all been tightly penned into one area. The farmer was waiting to greet them.

'There you are,' he said. 'I was wondering if you were coming down. I want you to give me a hand to separate the ewes from the lambs. I've got the shearer coming this afternoon and he'll expect to have everything ready when he arrives.'

Everyone helped to divide the ewes from the lambs while the farmer took each ewe in turn. He swiftly upended each one so that it was sitting upright and slightly leaning back against his legs.

'That's how you keep a ewe still while you are working on it. It stops her struggling,' he said. 'What I have got to do is to make sure that their feet are properly trimmed and that there is no foot rot. If somebody can hold that can of spray, I will squirt it onto any feet that are infected.'

While the ewes were being treated, their lambs kept up a constant bleating and the mothers seemed to be answering them.

'Do you think they are talking to each other?' asked Winston.

'Don't be daft,' replied Wayne. 'They can't speak!'

'What do you think they are doing then?' retorted Winston.

'I don't know,' said Wayne. 'Just making a noise because they are frightened, I expect.'

The farmer turned to Wayne and said: 'You wait and see. When they are all put back into the field this afternoon, the lambs will soon find their mothers. They are communicating with each other all the time. Marvellous really, isn't it?'

'I think that animals are just like humans,' said Carol.

'Perhaps we humans are just like animals,' replied the farmer.

Just then the farmer's wife joined them, carrying a large jug of squash and some freshly baked rolls. With the sheep all thoroughly checked and treated, it was time for lunch. As they were tucking in enthusiastically to their refreshments, the shearer arrived and began to set up his clippers.

It was a fascinating afternoon, watching the speed and skill with which the ewes were sheared. Some helped to guide the sheep through the pen, so that they were ready for shearing. Others gathered the fleeces as they were folded and tied. Touching the fleeces made their hands oily from the lanolin in the wool. A 'sheepy' smell filled the air.

When the shearing was finished and the sheep safely returned to graze, some of the animals needed to be fed. The cows were ready for milking, too, and so the farmer took everybody to watch what was happening.

'How have these animals got like this?' asked one of the boys. 'I mean, you don't find them like this in the wild, do you?'

'Breeding,' replied the farmer.

'How's that, then?' asked Winston, genuinely interested in what the farmer was saying.

'Well, to cut a long story short,' he replied, 'we humans have only been farmers for about 10 000 years. What we did was find some animals which were easy to domesticate and then, by carefully controlling their breeding, we've built up those things about them that we want to use. You take those sheep there, for instance. They've been specially bred for their meat. The ancient ancestors of these sheep were probably much longer in the leg, but that's no good if you want a good joint of roast lamb, is it?'

'I suppose not,' said Winston.

On the way back

The day had gone quickly and it was time to go; so they thanked the farmer for letting them spend the day at his farm and set off back to camp.

Suddenly Mr Payne halted. 'Hey, just a minute, stand still, quickly, look over there! Can you see that bird circling over the hedgerow?' he said.

They all looked in the direction in which he pointed.

'What is it?' asked Carol.

'It's a buzzard,' replied Mr Payne. 'Watch it for a while and see what it does.'

Suddenly the large bird dipped low in the sky and swooped down on a young rabbit.

'Did you see that?' said Mr Payne, obviously excited.

'That's its dinner for the day. It's probably got young ones that it's feeding.'

'I think that's really cruel,' said Dawn. 'I don't know why you find it so exciting.'

'That's nature, isn't it?' replied Wayne. 'Survival of the fittest and all that. I don't reckon that rabbit's very fit now, anyway!'

'Nature red in tooth and claw,' added Mr Jones. 'Isn't that right, Pat?' he asked, as he looked towards Miss Ridgewell, seeking approval.

'No, I don't agree with that,' she replied. 'They used to say it in the nineteenth century, following Darwin, but most biologists nowadays think that it's a one-sided view of nature.'

'Why's that?' replied Mr Jones, wanting to know why she did not agree with him.

'Well, I can think of plenty of examples of nature working in harmony. It's not all struggle for survival and blood. Of course things die, but if nothing died there could never be new life either. That seems to be the way it *has* to be.'

'What's Darwin?' asked Carol.

'Darwin isn't a thing. He was a person,' replied Miss Ridgewell. 'He suggested that living things evolved by a process he called *natural selection*,' she added. 'All the different species of plants and animals have developed from earlier, more primitive species'.

'Darwin wasn't the first to put forward the idea of evolution,' Miss Ridgewell continued. 'What he did was to suggest a way in which it could have happened, and then he collected a lot of evidence to support his theory. We'll be covering evolution in our lessons at school soon.'

As they continued on their way across the field to the farm, Leila felt that she wanted to ask Miss Ridgewell more about what she had been saying.

'Do you think it's true, Miss, about humans coming from apes and all that?' she asked. 'There are some people at the mosque, where I go, who have said that it isn't true'.

'That's right,' added Dawn. 'There are some people at my church who don't believe it either. They say that we all come from Adam.'

'That's because they're stupid,' interrupted Wayne. 'I reckon religion's just a load of fairy tales.'

'Just be quiet, Wayne,' said Dawn. 'We all know what you think. This is a serious discussion.'

'I didn't say it wasn't, Dawn,' Wayne replied.

'What do you think, Miss?' asked one of the other girls who had joined in the conversation.

'I think that humans and apes may have had a common ancestor; anyway, it's pretty certain life *has* developed over millions of years,' she replied, 'but I know some people think that the Bible teaches otherwise.'

'That's what I said,' interrupted Wayne. 'They believe in a load of fairy tales that aren't true.'

'I'm not sure I said that, did I?' she replied. 'Surely you can think of stories which say things that are true without the story itself having to be true, can't you?'

Things to do

Thinking about the story

1   What did Miss Ridgewell say Darwin suggested about the development of life on earth?

2   How did the farmer explain that farm animals had developed?

Points for discussion

1 In the story, Dawn thought that it was cruel when the buzzard killed the rabbit for food. People have often said that nature is cruel and evolution worked through this cruel struggle. Some people say that such a world could not have been created by a loving God. Do you agree with this view? Is nature only cruel? Can you think of examples of how good comes out of it?

2 Do you agree with Wayne that religion is based on 'fairy stories'? Why do you think stories have often been used to illustrate religious beliefs?

Creative responses

After reading the questions below, make a pictorial chart. On one side list the main character and the main event in each story, and on the other side write out the moral meaning or lesson of the story.

1 Read the story in Luke 15, 11–32 told by Jesus to show God as 'Father'. What is the meaning of this story?

2 Recall a story of a scientist (e.g. Newton and the apple, or the discovery of penicillin). What is the scientific truth put forward by your chosen story?

3 Choose a historical story (e.g. King Harold and the arrow, King Alfred and the cakes, or Robert the Bruce and the spider). What is the political or moral lesson of the story?

4 Dramatise in the classroom one of the above stories.

Topic work: Evolution

1 How did life begin on earth?

Earth is a planet teeming with life. Unlike the other planets in our solar system, earth developed in such a way that life was possible. The sheer number of different forms of life which this planet supports is staggering. As a 12 year old boy joked: 'If Noah had to take two of every species onto the Ark, he must have taken years rounding them all up!'

Living things range from very small micro-organisms which cannot be seen without the aid of a microscope, to gigantic organisms such as the redwood conifers of California, elephants, giraffes or whales (Figure 1).

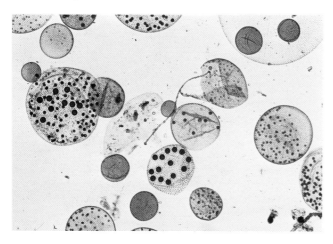

Figure 1 Micro-organisms and large animals

One question which has fascinated human beings for centuries is where all these forms of life came from – how did life begin on earth? Here are two possible answers; there are many more.

1 All living things were created in their present form, as we see them now: pine trees as pine trees, sheep as sheep, human beings as human beings. This is called '**fixity of species**'.
2 Somehow there has been a gradual *development* from one life form into the much more varied forms that we see now. This is called **evolution.**

2 Fixity of species

The view that all forms of life were created as they are now was the most widely held view before the middle of the nineteenth century. Aristotle had referred to the developments which take place *within* species, but the idea that one kind of plant or animal could develop into a completely different kind seemed to go against common sense. Animals of different species do not generally mate with other species, and we cannot see evolution taking place during our lifetime.

In the ancient Hebrew writings of the Book of Genesis the plants and animals are said to have been created 'according to their own kind'. Some people have interpreted this to imply the fixity of species. The phrase 'according to their own kind' can just as well be understood as expressing the orderliness of creation: dogs have puppies (and not baby rabbits), and cats produce kittens rather than something else.

3 Early ideas of evolution

Some people in ancient times and more recently thought that perhaps life evolved from earlier species. In the eighteenth century, a Frenchman called Lamarck developed a theory of evolution. He started from the fact that we do hand on some characteristics to our children. For instance, if you have brown eyes, then both your mother and your father must have brown eyes. Lamarck assumed that an animal might start off rather like a deer and, in trying to reach for the higher leaves on the trees, it slightly stretches its neck. When it mates, perhaps its young are born with slightly longer necks and then, as they reach even higher, their necks become even longer and they hand that on to their young. After

Figure 2 Giraffes stretch to reach the leaves at the top of trees

some generations we have a giraffe (Figure 2)! This is a different species of animal. Lamarck suggested that other animals have adapted to their surroundings or environment in similar ways.

That might look like a solution but there is a snag. There is no evidence that we hand on characteristics which have been gained in our own lifetime. The only characteristics that we hand on are those that we are *born with*. So, if you spent many years building up your muscles, you would not hand on your muscular body to your children. Lamarck's theory seems to have taken us up a blind alley.

4 Darwin's theory of natural selection

Many people believe that the theory of evolution was first suggested by Charles Darwin. This is not true: ideas of evolution had an earlier history. However, the earlier theories did not explain *how* life had evolved. Charles Darwin did. He suggested that evolution had occurred through a process of *natural selection*. Alfred Russel Wallace had hit upon the same idea whilst working on the tiny volcanic island of Ternate between New Guinea and Borneo.

Charles Darwin (1809–1882)

Charles Darwin (Figure 3) was born of wealthy parents. His father was a doctor in Shrewsbury and, when Charles left school, he was sent to Edinburgh University to study medicine. He disliked his studies so much that his father arranged for him to go to Cambridge with the idea that he would become a priest in the Church of England. While he was there,

he became interested in biology and geology, and he also enjoyed reading the books of William Paley.

Paley noted how all living things are wonderfully suited to their particular environment. For instance, fish are streamlined so that they can move easily through water. Paley said that this proves they have been designed for that purpose by God.

Figure 3 Charles Darwin in the last year of his life

However, as a result of a 5-year voyage around the world, Darwin began to see a way of explaining how all living things suited their environment. He took particular note of certain animals that he saw on the Galapagos Islands in the Pacific Ocean. He counted 13 different kinds of finch, and he was told that each of the islands had its own species of giant tortoise. People who lived there could tell which island the tortoises came from. Darwin wondered why, if God had created everything as it is now, he had bothered to create a special species just for one small island? It seemed to Darwin more likely that the various species were not fixed – or specially created – but that they had somehow *developed* from other species. At that time, however, he had no idea *how* that could have happened.

Putting the pieces of his theory together
Darwin's theory of evolution by natural selection
provides an interesting example of how scientific
theories may arise as a result of drawing together
ideas from different areas of knowledge. When
Darwin set out on his voyage round the world, he
already had two main ideas which later proved
important. Paley had shown how *all* species are
suited to their particular environment and Sir
Charles Lyell, a British geologist, had stressed that
the present shape of the earth was the result of
gradual changes over a very long period of time. On
his voyage, Darwin noticed the sheer *variety* and
abundance of species. He was also very interested in
the way that farmers and other animal breeders
gradually improve their stock. Animals *born with
special features* are selected and bred with other
animals with the same special features. In this way,
farmers gradually improve their stocks of animals.
For instance, racehorses which are good race
winners are mated together, normally producing
more good race winners. Horses which do not win
many races are excluded from the breeding stock.

Darwin realised that, if this process of selection

were to happen in nature, animals could change so
much over millions of years, that they would form a
new species. How, he wondered, could selection
occur naturally?

The missing piece
All these things were churning around in Darwin's
mind when, in 1838, he happened to read a book on
population and food supplies by the Reverend
Thomas Malthus. In the book, Malthus argued that
the human population naturally increases faster than
food supplies. This means that before long there is
not enough food available to feed the growing
population. The world would soon be overpopulated
if it were not for disasters such as famine and disease
which keep the population in check. Malthus
concluded that there must always be a *struggle for
survival.*

This gloomy prediction provided Darwin with the
missing piece of the jigsaw. He suddenly realised
that in all nature there is a struggle for survival. Only
those species most adapted to their environment
tend to survive. Those not suited to their
environment tend to be destroyed. At once, he could

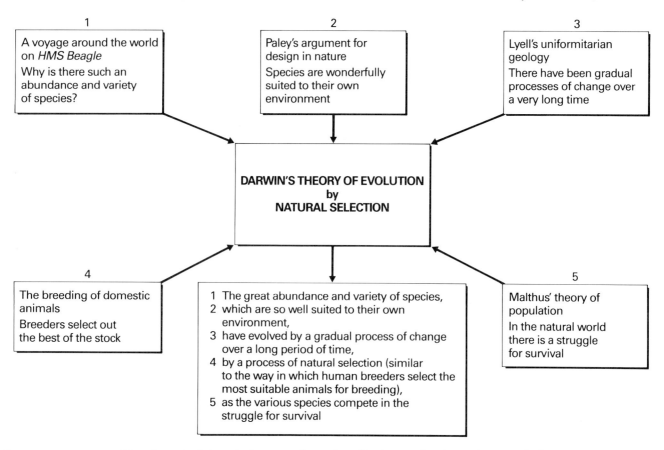

Figure 4 A summary of the ideas which Darwin drew together to form his theory of evolution by natural selection

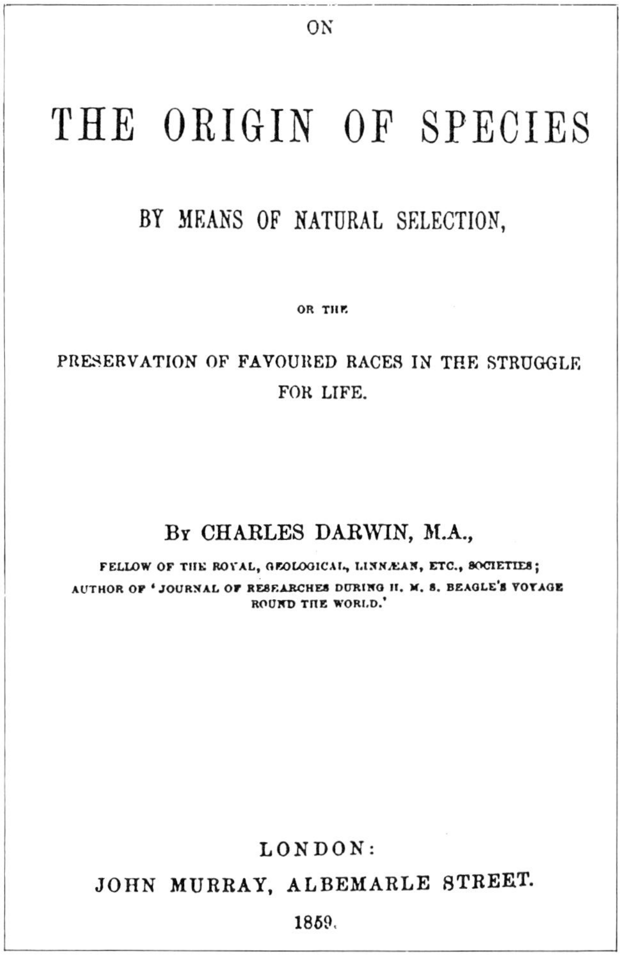

THE ORIGIN OF SPECIES

BY MEANS OF NATURAL SELECTION,

OR THE

PRESERVATION OF FAVOURED RACES IN THE STRUGGLE
FOR LIFE.

By CHARLES DARWIN, M.A.,

FELLOW OF THE ROYAL, GEOLOGICAL, LINNÆAN, ETC., SOCIETIES;
AUTHOR OF 'JOURNAL OF RESEARCHES DURING H. M. S. BEAGLE'S VOYAGE
ROUND THE WORLD.'

LONDON:
JOHN MURRAY, ALBEMARLE STREET.
1859.

Figure 5 The front page of Darwin's book

see how the various ideas fitted together. 'At last', he exclaimed, 'I have a theory to work with!' – the theory of evolution through natural selection (Figure 4).

5 Opposition to Darwin

Darwin wrote a book on this theory of natural selection, which was published in 1859 (Figure 5). Immediately there was a protest from some parts of the church. The Bishop of Oxford and some others condemned it as directly contrary to the Bible story of creation. They said that it was nothing less than a challenge to God's message to us about the way that he created the world. They saw Darwin's theory as an attack on belief in God and belief in the dignity of man. To some people, Darwin seemed to be 'making a monkey out of man' and destroying the basis of religion.

Other parts of the church, notably the

evangelicals, were silent. They took both the Bible and science so seriously that they did not wish to interpret the Bible in ways which went against scientific evidence.

To understand something of this controversy, arrange yourelves into two parties: one party for the Bishop of Oxford and one for Darwin.

The Oxford set could base their points of view on five propositions.

1 God made the world.
2 God took 6 days.
3 God made plants and animals 'according to their kind' or species.
4 God made man in a special way.
5 Whatever the Bible says is true.

The Darwin set could base their theory on five propositions.

1 All things evolved.
2 Evolution took millions of years.
3 Different species evolved from other animals.
4 Man evolved like any other animal.
5 Scientific evidence can prove these points.

It should be noted, however, that some non-Christians accept evolution and some reject evolution on scientific grounds; similarly, some Christians reject evolution, others reject evolution on biblical grounds and yet others accept evolution.

Today, a number of scientists challenge various parts of Darwin's theory, as we shall see later. Evolution was probably a lot more complicated than Darwin's idea of natural selection suggests (Figure 6).

Purpose and design in nature: a problem raised by Darwin's theory

Darwin said: 'There seems to be no more design in the variability of organic beings and in the action of natural selection than in the course which the wind blows.'

Darwin's theory of natural selection challenged the beliefs of Paley, and many others, who argued that God had *specially designed* all life forms. Let us see how the problem arose by considering the giant tortoise of Galapagos.

Look at the two tortoises shown in Figure 7. They are from different islands in the Galapagos. The tortoises are also different. The saddle-backed tortoise shown in Figure 7(a) has a differently shaped shell around its neck, and a longer and

THE
LONDON SKETCH BOOK.

PROF. DARWIN.

This is the ape of form.
Love's Labor Lost, act 5, scene 2.

Some four or five descents since.
All's Well that Ends Well, act 3, sc. 7.

Figure 6 Darwin's ideas were often made fun of by cartoonists

(a)

(b)

Figure 7 Two giant tortoises from different islands in the Galapagos

thinner neck. Each species of tortoise is specially suited to its particular island environment.

Now let us see how Darwin explained the difference and compare it with what we can imagine that Paley might have said if he had seen the two kinds of tortoise (Figure 8).

Perhaps you can try to consider another example for yourself. Darwin also noticed many different forms of finch (Figure 9). Their beaks were suited to different forms of feeding. How do you think Darwin explained the differences? Compare that with how you think Paley might have explained them.

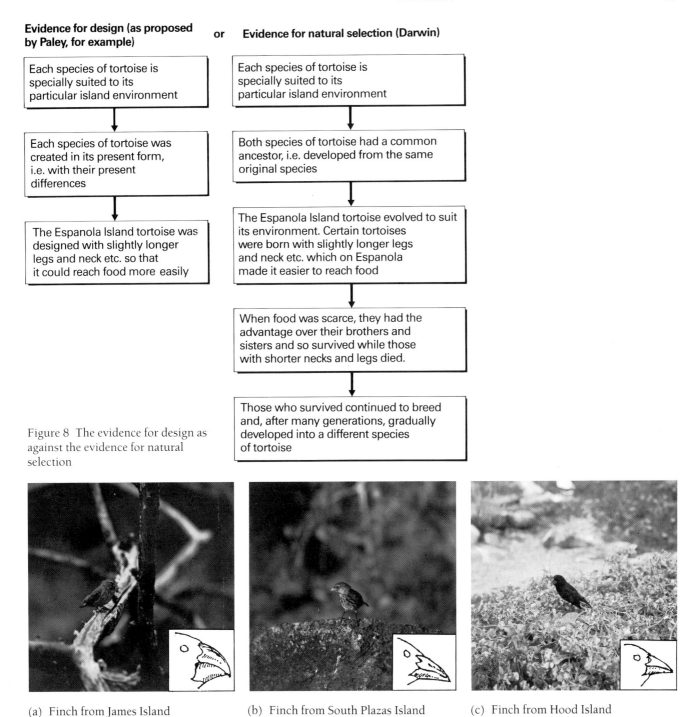

Evidence for design (as proposed by Paley, for example)   or   **Evidence for natural selection (Darwin)**

Each species of tortoise is specially suited to its particular island environment

Each species of tortoise was created in its present form, i.e. with their present differences

The Espanola Island tortoise was designed with slightly longer legs and neck etc. so that it could reach food more easily

Each species of tortoise is specially suited to its particular island environment

Both species of tortoise had a common ancestor, i.e. developed from the same original species

The Espanola Island tortoise evolved to suit its environment. Certain tortoises were born with slightly longer legs and neck etc. which on Espanola made it easier to reach food

When food was scarce, they had the advantage over their brothers and sisters and so survived while those with shorter necks and legs died.

Those who survived continued to breed and, after many generations, gradually developed into a different species of tortoise

Figure 8 The evidence for design as against the evidence for natural selection

(a) Finch from James Island   (b) Finch from South Plazas Island   (c) Finch from Hood Island

Figure 9 Darwin found differences in the beaks of finches living on different islands in the Galapagos

6 Science and theologians

Darwin got into trouble for the idea of 'evolving man' published in his book *On The Origin of Species*. Galileo had earlier got into trouble for the idea of 'moving earth' presented in his book *The Two Principal Systems of the World*.

The arguments against Darwin were very similar to those against Galileo, although hundreds of years had gone by since Galileo.

7 The Bible and evolution

Let us look at the Darwin conflict. Were the people who opposed him right in thinking that the Bible contradicts evolution?

In Unit 1, we studied Genesis 1. Look back at this and also study the following; see whether Genesis contradicts Darwin.

Genesis 1, 26, 27, 31

26 Then God said let us make man in our image. . . .

27 So God created man . . . male and female he created them.

31 . . . And there was evening and there was morning, a sixth day.

You probably know that there is another creation story in the Bible. Read this in Genesis 2.

Genesis 2, 7, 19

 7 Then God formed man from the dust of the ground and breathed into his nose the breath of life . . . .

19 Out of the ground the Lord God formed every beast . . . and brought them to the man to see what he would call them.

The two stories are not at all the same. They are different in style and also in what they say. Compare the two stories for yourself and make a list of the differences.

Nowadays we know from study of the Bible and other ancient writings that the writings in Genesis were originally passed on by word of mouth from generation to generation. At some later stage in their history they were gathered together. When there was more than one story of the same event, e.g. about creation, the different stories were included. Even though the stories focused on different details, the message was the same: the world was created by God for a purpose, God is good, and the world that He made is good.

The Bishop of Oxford (at the time of Darwin), modern-day creationists and evolutionists – whether Jews, Christians or Muslims – could all understand the message of these stories. They would not agree, however, on the picture of the world which forms the background to the stories.

When the stories of creation were written down, the ideas of the time, and the ideas of the readers or hearers, became part of the stories. In those days, people had a picture of the world that fitted the observed facts. The earth seemed flat; the sky seemed

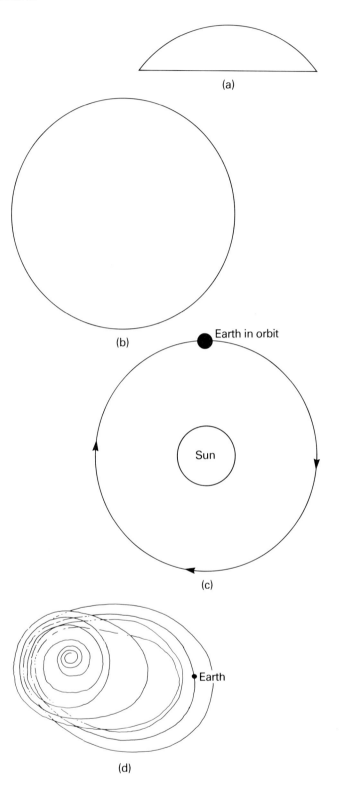

Figure 10 How a religious person might see changing views of the cosmos: (a) God made the world as a flat earth with a moving sun; (b) God made the world as a round earth with the earth at the centre of the universe; (c) God made the planet earth to move around the sun; (d) God made the whole galaxy to move in space

like an arch; the sun, moon and stars moved across the sky. This world picture is the framework of the stories. Today we have a different picture of the world. Can the message of the Bible survive this change in our picture of the world (Figure 10)?

A possible answer to the evolution controversy between Darwin and the Bible is as follows.

The meaning of the stories in Genesis tends to get lost, once you start to ask modern scientific questions about them. Equally, it is absurd to imagine that the writers of Genesis 1 and Genesis 2 could share our ideas about the earth as a globe going round the sun when all their experience was that the earth is flat with a dome for the sky. The authors maintained their own way of thinking when they wrote about God – whether as a potter (Genesis 2, 7), as a garment maker (Genesis 3, 21) or as a person 'taking the afternoon air' (Genesis 3, 8).

To take this story as a statement of a scientific process is like telling a scientist that he is unscientific because he talks about 'getting up at sunrise'.

Now read the following conversation carefully. Try to sort out what each speaker believes. Copy Figure 11 and put the name of each speaker in the box best suited to their belief (one name may be put in more than one box but, if you do this, draw a line between them).

Sarah: Of course, life has evolved from simple to more complex forms; there's no need to bring a non-existent God into it.

Brian: But there's no *proof* about evolution. In fact, something which hasn't evolved must have started life off and imprinted itself in the world to account for the variety of forms of life, but of course I wouldn't talk about a God.

James: Brian's right about evolution, and the Bible says that God created it all, and separate species.

Ann: No, I think that evolution's got a lot going for it and why shouldn't God create by means of evolution?

Evolutionist (A person who believes in evolution)	Religious (A person who believes in (a) God)
Creationist (A person who believes in the creation)	Atheist (A person who does not believe in (a) God)

Figure 11

5

Discovering a hidden pattern

A problem with words

THE campers enjoyed their evening discussions. Sometimes they were planned, but sometimes they 'just happened' as everyone gathered round the camp-fire.

The previous evening one of the topics of conversation had been their lucky escape from the storm. The discovery of the tree had made them realise how close they had come to being struck themselves.

The following morning they were still talking about it over breakfast.

'I reckon it was a miracle we weren't caught by that storm,' said Wayne.

'*You*? Believing in miracles?' gasped one of the other boys.

'Oh, I don't know about things like that,' replied Wayne. 'Perhaps I used the wrong word.'

'Words are funny things,' said Mr Jones, picking up Wayne's comments. 'Some things are easy to describe with words and others are very difficult.'

'You'll be learning some new words today,' said Miss Ridgewell. 'I hope you're all ready for a hard day's work.'

Studying the ecosystem

'We're off on a field study,' she replied. 'We are going into the wood to see whether we can discover something about the ecosystem operating there. What we are going to do is to dig down into the layers of soil to see whether we can find the detritus feeders which live there.'

'What are they, Miss?' asked Wayne.

'Well, they are things like worms and ants,' she replied. 'They are an important part of the ecosystem.'

'All these "long words" . . .,' said Carol. 'I hope it's not going to be too technical.'

'The idea is not that difficult at all,' replied Miss Ridgewell, as she produced a chart which she had brought with her. 'You understand what

I mean by a community, don't you? An ecosystem is the community of all the plants and animals and non-living things which inhabit a particular area, in this case the wood. Everything has its own part to play in keeping the ecosystem going.'

'How do you mean, Miss?' asked Carol, still unsure that she was understanding all she was being told.

'There we are,' said Miss Ridgewell, pointing to her chart. 'There you can see the *detritus feeders* which I mentioned earlier. They have a very important job to do because they break down the remains of plants and animals and their waste products into smaller particles so that the so-called *decomposers* such as fungi and bacteria can turn them into the valuable nutrients needed to support plant life. Plants are eaten by the herbivores (or plant eaters) and in the ecosystem they are called *primary consumers.* These are then eaten by the carnivores (or meat eaters) which are called *secondary consumers.* When the plants and herbivores and carnivores die, they decay and their remains are then returned to the soil for the detritus feeders to decompose again.'

'So the buzzard that we saw the other day was a secondary consumer eating the rabbit which was a primary consumer?' enquired Carol.

'Exactly,' replied Miss Ridgewell.

There was plenty of enthusiasm for the task set before them. The wood, which had seemed threatening when they had walked through it in the dark, took on new meaning. They had noticed very little evidence of activity when they had walked quickly by, but now they could see that the wood was teeming with life and movement.

Keeping itself going

'It's strange how it all keeps itself going, without realising it,' said Winston.

'What do you mean?' asked Miss Ridgewell.

'Well, you take that ant we were watching a little while ago,' he replied. 'It doesn't know that it is helping to keep the whole wood alive does it? Perhaps it only thinks that it is keeping itself alive.'

'Ants don't think,' interrupted Wayne.

'How do you know?' replied Winston, annoyed at being interrupted.

'They must be conscious in some way or another which we cannot appreciate,' said Mr Payne. 'Don't you think it's remarkable that the natural world keeps itself in balance, Wayne?'

'That's why religious people are wrong,' said Wayne, 'because they believe that God keeps everything going and you have now told us that nature keeps itself going.'

'Yes, what about that?' asked Dawn, for once finding herself in agreement with Wayne. She even found herself smiling at him, and he returned the smile.

'It isn't a question of either/or,' replied Mr Payne. 'Religious people believe that God makes nature like it is.'

'But people say that science has shown that God doesn't exist at all, Sir,' replied Winston.

'That's right,' said Carol. 'My dad said that in the past religion has always fought against science and in the end it has always lost. You don't know what to believe nowadays. Science says there isn't a God and religion says there is.'

'So what you are saying is that you have to choose between one or the other,' asked Mr Payne. 'You couldn't be a scientist and also a believer in religion?'

'Of course you couldn't,' said Wayne, emphatically.

'At any rate you couldn't be a good scientist.'

'Why's that, Wayne?' asked Miss Ridgewell.

'Well, like I said, science shows that the world runs itself; so where does God fit in? He's redundant isn't he – out of work,' replied Wayne.

'He could have created it in the first place,' said Carol.

'Scientists don't know everything, do they?' said one of the boys. 'Perhaps God is doing the bits we don't understand.'

'What happens when we do understand, then?' replied Wayne.

'You're right, Wayne,' said Mr Payne. 'That's tended to happen in the past. God was gradually pushed out as our knowledge grew – that kind of God has been called the *God of the gaps.*'

'Do you believe in that kind of God? After all you're the RE teacher,' said Dawn.

'No, I don't see it like that,' he replied. 'I believe that God created the world so that it would work in an orderly way and so that human beings could enjoy finding out about it all.'

'So you mean that's where science fits in, Sir? Could you sort of say that God made science?'

'Yes, for someone who believes in God, there would be no science and no scientists without God. Everything depends on God for its existence. In other words the world doesn't have to exist and one day God might bring it to an end.'

'Surely scientists don't believe that sort of thing,' interrupted Wayne looking at the science teacher, Miss Ridgewell.

'There are plenty of scientists who are religious people. I think they would agree with that,' she replied. 'I'm sure they would say there is something rather wonderful about the whole of creation. Of course, the world works according to natural processes just like the ecosystem we have been studying today but, even though we can describe what is happening, there is something rather mysterious and gripping about the whole of creation. The more we know, the more wonderful it seems. At least that's what I've found but I can only speak for myself.'

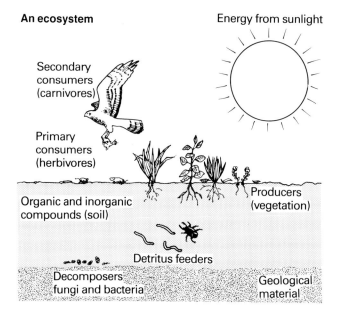

An ecosystem

Energy from sunlight

Secondary consumers (carnivores)

Primary consumers (herbivores)

Organic and inorganic compounds (soil)

Producers (vegetation)

Detritus feeders

Decomposers fungi and bacteria

Geological material

Things to do

Thinking about the story

1 Why do some people, such as Wayne, think that you cannot be a good scientist and also have religious beliefs? Do you agree with them?

2 Do you ever wonder whether small life-forms, such as ants, can 'think' in some way or another? How can you explain their ability to work together?

Creative responses

1 In the story, Miss Ridgewell used a chart to show how an ecosystem works. The picture shows the kind of chart she would have used. Copy the chart carefully and then explain it in your own words. Imagine that you are trying to explain the idea of an ecosystem to a friend who has not heard of it.

2 Try to think of something which is very hard to describe in straightforward language. Perhaps you might think of falling in love,

the joy of seeing a newborn baby or the feelings that you had when you experienced something which really stirred you, perhaps some music or something you experienced in nature or a religious moment. Then see whether you can express what you are thinking about in language using either poem or prose form. Try to use the language as imaginatively as you can so that another person reading it would capture something of what you felt at the time.

Topic work: 'The world runs itelf; so where does God fit in?'

The title of the topic work for this unit is one of the comments which Wayne makes in the story. In thinking about an answer to it, much depends on the picture of God which people have.

1 Different pictures of God

What we think about this question of miracles depends on what we think about God and how God is related to the world.

Whether you believe in God or not, what is the picture of God which you have?

Pictures of 'God'

1 A God who is like a slot machine whom you can approach with your coin and get out what you want.
2 A God who presides over a very boring harp-playing assembly.
3 A God who is like a father.
4 A God who gives obvious signs of his presence which cannot be contradicted.
5 A God who is not bothered about the universe that he started.
6 A God who is the conclusion of a mathematical theorem.
7 A God who is a sort of celestial policeman.
8 A God either who is evil or who is powerless to do anything about it.
9 A God who does not give people independence, freedom and self-respect.
10 A God who is visible, whom scientists should have seen in a telescope or microscope, or whom astronauts should have found.

Only one of these pictures of God is how a genuinely religious person would be likely to think of God. Which one?

If you have grown up within a religious tradition yourself, perhaps you could explain to others in your class what your religion teaches about God. You may be able to compare ideas of God from different religions.

Figure 1

2 God and the world

In the story, the conversation discussed how God might be related to the world.

Religious people see the relationship between God and the world in different ways. In this section we shall look at *three* of these: **Deism**, **Theism** and **Pantheism**. These are technical terms which sound quite a mouthful, but they are useful and worth learning because they sum up something special, just as in science technical terms such as ecosystem are used.

Deism (from the Latin word 'deus' meaning God) Some people believe that God started the world in such a way that it could go on by itself. God therefore plays no further part in the running of the world. This is close to the idea which Aristotle had of God as the First Mover (see Unit 1, page 9).

The idea that God is the great designer (Figure 2(a)) of the world goes back further than Paley. Figure 2(a) comes from a thirteenth-century French Bible. Can you describe what you see in the picture?

Look at Figure 2(b). Each part of the eye seems to be made for a purpose. Together the various parts enable us to see.

Deism became popular in the eighteenth century.

(a)

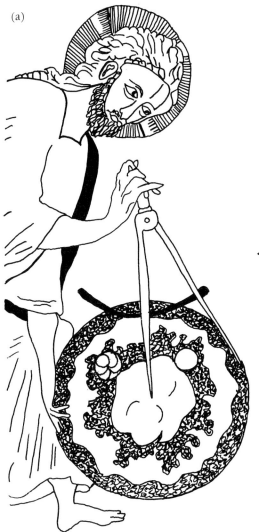

(b)

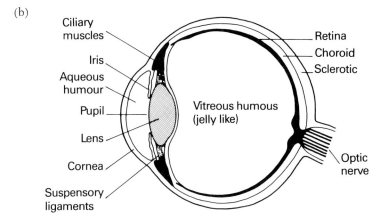

◀ Figure 2 (a) God the great designer; (b) the human eye

Figure 3 'God is a kind of watchmaker'

This was when people were coming more and more to think of the world as though it were some kind of vast machine (Figure 3) which, once made, would keep on going on its own.

Theism (from the Greek word 'theos' meaning God) This is like Deism in that people believe that God made the world in the beginning, but it is also different. In this view, God goes on relating to the world. God is involved in the running of it and can alter it.

This is how Peter Hodgson, a scientist in Oxford, puts it: 'God not only created the universe; He keeps it continually in being. The universe depends on God not as the watch depends on the watchmaker, but as the song depends on the singer. The dependence is continual and total. Without this sustaining power, all would lapse into nothingness.'

Can you think of another metaphor like the song depends on the singer (Figure 4), or the clock depends on the clockmaker?

Pantheism (from the Greek words 'pan' meaning everything, all and 'theos' meaning God) Pantheism means everything is God, or God is everything. God is the same as the world.

This is an ancient way of looking at the world and God. It does not make a clear distinction between them. People who have mystical experiences often come close to this view because they feel themselves to be part of their physical surroundings and part of God at the same time (Figure 5).

Figure 4 The song depends on the singer

Figure 5 The wonder of nature can inspire devotion

Poets and artists sometimes believe something like this. In the following poem in which Shelley writes about the death of a friend, which words or phrases suggest this view?

From Mourn not for Adonais *by P. B. Shelley*

He is made one with Nature: there is heard
His voice in all her music, from the moan
Of thunder, to the song of night's sweet bird;
He is a presence to be felt and known
In darkness and in light, from herb and stone,
Spreading itself where'er that Power may move
Which has withdrawn his being to its own;
Which wields the world with never-wearied love,
Sustains it from beneath, and kindles it above.

3 Miracles in a 'scientific' world

At the beginning of the story, Wayne said that their escape from the storm had been a 'miracle'. This is a word which is used, particularly in a religious sense. People who believe in God often also believe that miracles can happen. But does not science explain how things happen? So why bring God into it?

Imagine that you have a friend who is desperately ill in hospital. He or she is suffering from an incurable disease and the doctors are unable to help. It seems that nothing can be done. But then, quite suddenly and without any explanation, your friend begins to recover and in a short while is able to leave hospital, completely cured.

- How would you explain what had happened?
- Would you say it was a miracle?
- Are there any other possible explanations for events like that?
- Would it make any difference to your answer if you knew that people had been praying for your friend's recovery?
- Do you think that it *is* possible to prove that miracles occur?
- Can you give reasons for your answer?

Belief in miracles and science
People today often find it difficult to believe in miracles and many of the events which would have been called miracles in a former age can now be given a 'scientific' explanation. In other ages, surprise events were seen as special acts of God.

The word 'miracle' itself comes from the Latin word 'mirabilis' which means 'wonderful'; so it indicates something which excites wonder. In this sense, many events which can be given a very full scientific explanation are often called miracles, events such as the birth of a baby, a heart transplant, or things which happen to turn out all right through coincidence (e.g. someone who did not catch a train which crashed). Wayne used the word miracle this way in the story.

But religious people see a miracle as something more than just a wonderful thing which chanced to happen. It is an indication of God's activity in the world in a different way to that in which He acts in the normal running of the world.

Theology therefore is concerned not with explaining how miracles work but with the historical evidence and the credibility of the witnesses to a miracle. The Bible

1   states that miracles can happen,
2   gives evidence that some miracles actually did happen and
3   says that these miracles are a sign of God's action.

Looking at the laws of nature

It is important to remember that the word **law** is a metaphor used in science to express how things tend to happen in the same way, given the same conditions. Technology is when people make certain laws dominate certain other laws. Thus, for example, when an aeroplane is in the air or when a rocket is launched (Figures 6 and 7), the 'laws' of aerodynamics overcome the law of gravity. A law of nature is a summary of scientific evidence and tells us what we can expect to happen.

Seen from a scientific point of view, when the unexpected happens, we look for some outside cause which accounts for this exception to the law. When we run out of 'outside causes' and can find no known explanation, we start to re-examine the law to see whether we have really summarised all the evidence.

Look at these sentences. They are all explaining the same incident. The room was filled with light

* because the electric bulb gave off light,
* because Jane put the switch on,
* because Jonathan was afraid of the dark.

The first explanation is just the mechanics of it. The second brings in the human agent. The third shows the purpose. These three explanations can

Hold a wet rubber ball between your finger and thumb. Spin the ball. Water will shoot out around the centre of the ball because this part of the ball is moving the fastest.

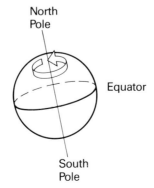

Similarly, as the earth spins on its axis, the surface of the earth moves fastest at the equator. This is why rockets are often launched from sites near the equator (for example, Cape Canavarel).

North Pole

Equator

South Pole

Figure 6 The rotation of the earth assists the launch of rockets

Figure 7 The launch of a rocket

John 9, 1, 6–12 (Anchor Bible)

1 Now as he walked along, he saw a man who had been blind from birth . . . .

6 Jesus spat on the ground, made mud with his saliva, and smeared the man's eyes with the mud.

7 Then Jesus told him, 'Go, wash in the pool of Siloam.' And so he went off and washed, and he came back able to see.

Figure 8 The pool of Siloam

each stand on their own and they do not need any other. But they are all true and do not contradict each other. A completely adequate explanation would give all three. When religious people insist on the fact that such and such an event is a miracle, they mean that God happened to be involved in it as well as the other factors.

This raises an important question. How do we know? Why do people ever believe that God is involved in an event? The early Christians saw remarkable things and recorded them in the gospels which they wrote, but were they mistaken?

4 Looking at the evidence

Let us look at one of the healing miracles recorded of Jesus. Do you think that it is likely or unlikely that Jesus healed the person?

8 Now his neighbours and the people who had been accustomed to see him begging began to ask, 'Isn't this the fellow who used to sit and beg?'

9 Some were claiming that it was he; others maintained that it was not, but just someone who looked like him. He himself said, 'I'm the one all right.'

10 So they said to him, 'How were your eyes opened?'

11 He answered, 'That man they call Jesus made mud and smeared it on my eyes, telling me to go to Siloam and wash. When I did go and wash, I got my sight.'

12 'Where is he?' they asked. 'I have no idea,' he replied.

Spittle (saliva) was believed to have medical qualities. Jesus therefore chose to use something which would help the blind person to have faith that he would see. He was also told to wash in Jerusalem's regular water supply. The miracle surprised people all the more because the man had been *born* blind. John tells us that there was a thorough investigation of the miracle at the time. The authorities in Jerusalem made frantic efforts to try to prove that there was an ordinary explanation.

Read the full text in John 9, 1–41. Then answer the following questions.

- Who were the main witnesses of this wonder?
- What did they think about this miracle?
- Do you think that John who included this in his gospel believed the witnesses?
- Compare the reactions of the man born blind (John 9, 15, 24–33) and of the Pharisees (John 9, 16, 33–34; Mark 4, 36–41). What was the sign or meaning of this miracle?

The stilling of the storm

Let us consider another miracle of Jesus. Look it up in Mark 4, 35–41.

The Lake of Galilee is well known for the sudden dangerous storms which happen without warning. Galilean fishermen, then as now, had to be very skilful sailors, but even so sometimes people drowned. Read the passage carefully. It is quite a graphic piece of writing. It perhaps reads like the account of an eye witness.

- Which words or phrases suggest that someone was reporting a real incident?
- Who were the witnesses?
- What did they think about this miracle?
- Do you think that Mark believed these witnesses?

Today we can try to make guesses about or to explain this story.

1 *Coincidence* The event actually happened, but Jesus was not the cause of the calming of the storm. That was just coincidence; Jesus happened to say the words 'Be still' and the storm by chance died down at that moment.
2 *Jesus had special power* The event actually happened as stated. Jesus really did have power over nature, so that 'even the winds and waves obeyed him'.
3 *Symbolic, not real* The event never actually happened. It was made up by the early Christians to teach others about the amazing difference in a person's life that following Jesus can make. The sea represents the difficulties of life; the storm

Figure 9 A storm on Lake Galilee

represents the chaos when things go wrong and a person looks like being submerged (whatever the cause, e.g. illness, someone leaving home, or failing an examination); the boat represents the Church. Thus the story says that Jesus is Lord as in Psalm 89, 8, 9.

> *Psalm 89, verses 8 and 9*
> 'O Lord God of hosts,
> who is a strong
> Lord like unto this?'
> Thou rulest the raging of the sea
> When the waves thereof arise thou stillest them.

- Which of these three explanations do you think is the most likely? Say why you think so.
- How could explanation 3 be altered to make it fit in with either explanation 1 or explanation 2?

Whether a person believes that such miracles happen depends on many factors other than scientific ones. History can establish that an event happened. Science however, cannot say that it could not have happened because science can find no natural explanation.

Figure 10 The altar at Lourdes showing the crutches left behind by people who have been healed

If 'unexpected' things keep on happening around a person (such as Jesus) or around a particular place (e.g. St Thomas' tomb in Canterbury), then people see the miracle as a religious sign: a sign that God is the cause of the unexpected to show that this person or place has God's special message or blessing.

There are some religious people, however, who would believe anything and would accept a miracle on the flimsiest of evidence. Great care is needed in weighing this up.

Nowadays at the pilgrim centre in Lourdes (Figure 10) there is a bureau of medical doctors to check on every alleged miracle before, after and at regular intervals afterwards. Some of the bureau are non-Catholics, some agnostics (people who believe that we can know nothing about God) and some atheists (people who do not believe in God). They look at each case rigorously to verify that the person was medically sick, that there was a medical change and recovery, that this recovery was sustained and permanent, and that they could not find any known medical explanation for the cure.

Discuss this comment: 'The arrogant scientist and the gullible religious person do not use their common sense.' Do you agree?

5 Miracles in the Bible

Jewish Scripture speaks of natural events and extraordinary events, as the 'wonderful' works of God. The idea is that God made the world, continues to keep it going and cares for people in a special way. So things such as a beautiful sunset, the birth of a child and the escape of the Hebrews across the Red Sea are all seen as wonders brought about by God who loves and cares for us. They are signs of God's providence.

The Christian Gospels see things in the same way. They were originally written in Greek and they use three different words to describe Jesus' miracles.

1 **Dunameis**   meaning   mighty acts.
2 **Semeia**   meaning   signs.
3 **Terata**   meaning   wonders.

The word **terata** is the word closest to our word miracle which comes from the Latin word mirabilis meaning wonderful.

The word **dunameis** is close to the English words dynamite, dynamo and dynamic. What do these words mean? What do they have in common?

The word **semeia** gives a clue to what the writers of the gospels wanted most to say about the miracles of Jesus. They were *signs* or pointers to God's character and purpose.

The gospels want to show that God was doing something unique and new in Jesus. So people should listen to his message. Jesus was not wanting to get a reputation as a wonder worker, nor was he trying to overthrow the natural way in which the world works. He was using the miracles that he performed to direct people's minds to God. The unusualness of the miracle could pull people up short. It could encourage them to think about a Power greater than that of nature. It could also teach them about God's attitude to the nasty side of life.

Christians who have thought deeply about the miracles of Jesus see them as teaching at least four things.

1 God does not like people to suffer any more than we do.
2 God is in control. Suffering from natural causes happens only because God allows it to happen.
3 When its purpose is achieved, God will put an end to suffering and all will be well. Good will come out of evil.
4 The miracles around Jesus are a sign that Jesus is a messenger from God and so his teaching is also a message from God.

The resurrection of Jesus

The New Testament records the greatest miracle as the resurrection of Jesus in the words 'He is alive; he is risen'.

Applying all our discussion so far to this, we can summarise the gospel story as follows.

The apostles said that they were witnesses to the facts that Jesus was alive and lived among them, and that Jesus died and was buried. After his death, quite unexpectedly, the same man was alive and was seen to be alive (not a ghost) but with the very body which had been crucified. They did not explain how it happened, but only the fact that it did happen. They witnessed that he had spoken with them and eaten meals with them.

The apostles say that such a happening could only have been brought about by God. Because God raised Jesus from the dead, people ought to listen to the things that Jesus taught.

Here then we have an example of an extraordinary event, vouched for by actual eye witnesses, which causes astonishment and which was seen as a sign that the teaching of Jesus must have been from God.

6
Camp inspection

All is not well

THE next day after breakfast Mr Jones announced that the headmaster would be arriving at the camp mid-morning. 'That doesn't give us much time to get the place tidy before he arrives,' he added.

The campers had a hurried breakfast and then set about tidying up the field. It had become much more untidy than they had noticed. The tents were also in a state of confusion, with odd socks and other garments turning up in the most unexpected places. It took a full hour to restore things to what they had been on the day that they had arrived. Even then the odd piece of litter could still be seen.

Most of the pupils liked the headmaster and were pleased that he had come to visit, but Dawn was not one of them. When he was far enough away not to hear, she said, 'He thinks he's like God, walking around inspecting everything as though he owned the place.'

'You mustn't say that,' replied Mr Payne, who had heard what she had said.

Winston, thinking about what Dawn had said, remarked, 'You imagine what God would think if *he* visited the earth and inspected it.'

'What do you mean, Winston?' asked Mr Payne.

'Well,' he replied, 'I'm very interested in animals and plants and things, and on television we are always being told that man is threatening life on earth. We have polluted the rivers, seas, and air and we have hunted some animals so that they are almost extinct. On one programme recently, they were saying how we are cutting down all the rain forests, which will threaten oxygen supplies.'

'I believe that is true as well,' said Leila. 'I was reading a book the other day about Islamic science and it was saying that in my religion we have been taught to be more careful how we treat the earth. It told a parable to show how humans have no right to treat life as they do.'

'I've just had an idea,' interrupted Mr Jones.

'When we gather at the camp-fire tonight, why don't you put on a play and act out your parable. Winston, you can help Leila. There are quite a few of you here from the drama club, so we ought to be able to do something without too much rehearsal. The headmaster is staying overnight and he would enjoy that.'

Preparing the plays

Leila began by explaining that it was a very old Islamic story. Islam had always been very interested in the study of nature and in how people are connected to the rest of the world. The story pictured an argument between human beings and all the other animals. The animals wanted to know why humans have the right to treat animals in the way that they do. In reply the humans gave many reasons including their greater intelligence and their ability to invent many clever things. The animals were not satisfied with any of the answers.

'Wayne will enjoy that,' said Winston, sarcastically. 'He loves stories about animals talking.'

After they had carefully worked out how they could turn Leila's story into a short play, Mr Jones went on to think what could be done with Winston's idea that God visited the earth. As it happened, Winston was very good at acting and had taken the leading part in school plays before. Mr Jones suggested that Winston should take the main part, which was God. The others would take on a number of different roles, each time showing a scene in which God comes face to face with human beings whose actions threaten the balance of life.

In the play the actors were to show the different kinds of attitude that people have to the earth. Sometimes those who were causing the damage did not realise what they were doing. Many were genuinely sorry. Others, however, did not care, so long as they got from the earth what they wanted. For instance, there were the lumberjacks who were cutting down the rain forests, so that grass could be grown to feed cattle, for sale as beef in America. There always seemed to be 'good' reasons for the destruction.

There was no time to prepare a written script, but they were quite used to performing short unrehearsed plays in their drama club at school.

Evening entertainment

It was decided to mark out a small stage between the camp-fire and the edge of the wood. The spectators sat on logs on one side of the fire while the plays were performed on the other, against the backcloth of the trees. Despite the odd mistake the actors managed to capture the spirit of Leila's story very well. The audience enjoyed it – especially the headmaster who continued to clap after everybody else had finished!

After the play, drinks were served, and then the actors prepared to perform a second play. This had been less well rehearsed than the first. Evening was drawing on and the flames from the fire silhouetted the trees against the darkness of the wood. The play was to open with a scene in which God was talking to the angels, who had visited earth to report on how man was caring for creation but, as it is not possible to portray God or to act God in a play, they had decided that Winston should play the part of 'Earth King'. So the angels reported to the Earth King about men working on earth. The Earth King was very disturbed to hear that some of man's activities were threatening the balance of life. He therefore decided to go and make an inspection of earth himself. He disguised himself as a wandering teacher and, when he reached the first group of humans, he was pleased to see how busy they all were. It was only when he realised that much of their energy was being spent on making weapons that he became angry.

'Why are you spending so much time and energy on that?' he demanded.

One of the humans pointed to another group not far away. 'Can you see those humans over there?' he asked. 'They are our enemies and are preparing weapons to destroy us. We have no choice but to do the same.'

The Earth King's anger turned to deepest sorrow. 'My children,' he said, 'that is not why God has given you the ability to reason and invent, so that you might spend time planning how to kill. Turn your efforts into helping one another and making earth into the kind of home that God planned for you when he created it.'

A turn of events

The play was going well and the audience were immersed in the action, but somehow Winston the King sensed that something was wrong. The audience were just *spectators,* looking on, but not really involved. They ought to be part of a play which is about human beings on earth. Suddenly Winston saw his chance. One of the boys had been so engrossed that he had been dropping sweet papers on the ground. Winston surprised everyone.

In his role as King he suddenly turned and pointed to the boy in the front row. 'You,' he shouted, 'you are just as guilty of carelessly treating your earthly home. All of you. Remember the way that you had turned that field of grass where you are camping into a pig sty? And how many of you have thrown away food when you know that so many of your brothers and sisters are starving? You have been wasteful. Instead of returning milk bottles, some of you found it easier to throw them in the rubbish. For all these things, and more, you *all* stand condemned. Each of you in your own way has treated the earth without care.'

Without warning the audience were no longer spectators but were instantly involved in what was happening in the play. The play was no longer about other people in other countries, who were threatening the homeliness of earth, but about their attitudes and the things that *they* were doing. It made them seem not much better than those who they had earlier wanted to blame. Many of the audience, including the teachers, felt uncomfortable by the turn of events. They were all quite embarrassed, especially the boy with his pile of sweet papers. His face looked bright red, and not just from the heat of the fire!

As the play neared its end, the actors faded one by one into the darkness of the woods. Only Winston was left on stage. Slowly he began to walk backwards and out of sight. As he disappeared from view he raised his voice and spoke more firmly and slowly – the words fading into the distance as he retired. 'You are all guilty. You are *all* guilty. I, the Earth King, have visited and inspected your camp and I have been troubled by what I have seen and what I have heard. Change your attitudes and change your ways or your earthly home will be taken from you, taken from you . . . taken from you . . . .'

The play had ended. Around the glowing embers of the fire, only the audience remained. But, as it had turned out, they now seemed to be the ones on stage. At least that is how it looked from among the trees where the actors were now gathered. The light of the fire was no longer on them. Instead it was spotlighting the audience, who had somehow become the real actors in the drama about what we are all doing to the planet earth.

Things to do

Thinking about the story

1 In Leila's story the animals put the humans through an interrogation. Are there any rules which mankind must obey?

2 In the play, why did the King suddenly become personal, and why did the boy's face turn red? Is a person's conscience an inner voice like the voice of the Earth King?

3 How did Winston make the spectators become participants? Is religion helpful in making a person into a participant in the world ecosystem?

Creative responses

1 Carefully read again the summary of the story which Leila told. Then try to imagine how they would have performed it as a play and the lines which they would have spoken. Write them in your book (it may be best to do it on rough paper first). Remember to use a variety of animals and try to think of some good arguments that they would have put to the humans and the replies which the humans would have made.

2 Modern scientists have released the energy inside the atom. A scientist says that nuclear power for making electricity 'is the cheapest, cleanest, safest way to produce energy without which the industrial world will come to a halt'. A politician says that a 'nuclear winter' after an atomic war will make our planet a place where man can hardly survive. What might the 'Earth King' in the story say to the scientist and the politician?

Topic work: Taking care of the world

1 The delicate balance of nature

Consider a pair of scales (Figure 1). The way in which they work is quite simple. They are so designed that the two arms of the scales are in perfect balance, pivoted on a central fulcrum. When a weight is placed on one of the trays, it will throw the scales out of balance until an equal weight is placed on the other tray. When the scale is motionless and equally balanced, we say that it is in 'equilibrium'.

This idea of 'balance' can be used to explain how the natural world maintains itself. The world of nature is in *dynamic* equilibrium. By this we mean that things are always happening and changing but yet keeping in balance. Just think of the continual changes which are taking place quite naturally in ecosystems. Every small change threatens the balance of nature – it is as though an extra load has been placed on one side of the scales. Yet, for the most part, the natural world works in such a way that the balance is continually restored. On some occasions the balance changes dramatically, a certain species, such as the dinosaurs, have become extinct.

Figure 2 Industrial pollution of the air

This helps to remind us that the balance of nature, like the balance of the scales, is very delicate and is easily upset.

Upsetting the balance
Human beings are threatening the balance of nature in many different ways. Look at the pictures in Figures 2–4. Each one provides an example of how we are overloading the balance and making it difficult for the natural world to cope.

The various gases which make up the earth's atmosphere have not always been the same. At the moment, about 70% is nitrogen, 21% oxygen and only 0.03% carbon dioxide. In the early stages of the earth's formation, there was less oxygen and more carbon dioxide. The balance was changed by small life-forms called algae. They increased the oxygen and reduced the carbon dioxide. This allowed more complicated life-forms to develop. Now, however, we are pumping vast quantities of carbon dioxide into the atmosphere (Figure 2). This will obviously upset the balance again, but we cannot be sure exactly what will happen. Some scientists predict that it may cause the earth's atmosphere to warm up

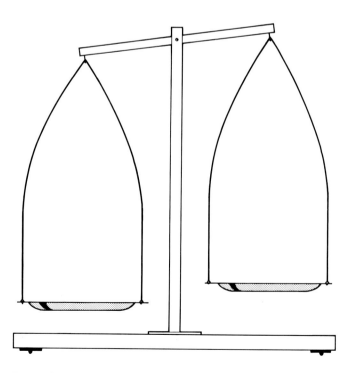

Figure 1

Figure 3 A polluted lake

which would melt the polar ice caps and make the sea levels rise all over the earth.

Large areas of the earth are covered by water, and their size is part of the fine balance of nature in our world. For example, we depend on the oceans not only for food but also because the minute life-forms which they contain supply the atmosphere with oxygen. We human beings have treated the seas as useful dumping grounds for raw sewage and industrial waste. These pollutants have already caused many lakes and rivers to become dead – no longer able to support plant and animal life (Figure 3).

Land surfaces are part of nature's balance. The land not only provides the space in which we live but is also our main supplier of food. Intensive farming (Figure 4) increases the food supply but this takes

Figure 4 Intensive farming

nutrients out of the soil. Concentrated fertilisers are used to keep the soil fertile when the natural process of humus making is destroyed. Some of the excess fertilisers are washed off the land by rain and find their way into rivers and seas where they pollute the water. Other chemicals are used in 'pest' control and these have killed fish and birds, upsetting the balance of the ecosystems in which they lived.

2 Spaceship earth

Have you ever thought that you would like to go on a space flight? Would it surprise you to know that you are already on one?

When scientists and engineers first began to design spacecraft, it was immediately obvious to them that they had to produce a miniature earth. There are certain things which the astronauts need just to keep them alive in space. Can you think what these basic needs are? Together they provide what is called the on-board life-support system.

As the first astronauts went further out into space, the earth receded into the distance and seemed to grow smaller. From out in space it appeared just like another spaceship. The astronauts had their on-board life-support system on their spaceship but spaceship earth had a similar system all around it, keeping its crew of plants and animals and human beings alive.

The early spacecraft carried only a limited supply of the basic requirements for life. As longer space journeys were planned, huge quantities of air and water etc. were needed. They had to devise a way of recycling waste products so that they could be used again. No spacecraft can carry unlimited supplies. This is still one of the biggest difficulties which face us if we are to undertake any very long journeys away from earth.

On earth the problem *has* already been solved in nature's balance. For millions of years, earth has journeyed through space and has managed to keep its inhabitants alive by recycling the basic requirements for life.

Recycling the water supply

Water is evaporated from the earth's surface and water system, forming clouds. As the clouds pass over the land, they drop the water as rain or snow. The water irrigates and replenishes the earth and provides fresh water for animals to drink. Much of it gradually drains back into the rivers and lakes and seas, thus completing the cycle (Figure 5).

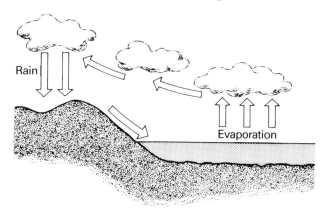

Figure 5 The water cycle

Recycling the air supply

Green plants are the main source of oxygen in the earth's atmosphere (Figure 6). Plants absorb carbon dioxide from the atmosphere through small openings called **stomata**. By a complicated chemical process called **photosynthesis** the sun's light converts the carbon dioxide into the nutrients required to make the plants grow. The process also produces oxygen which is required by animals. Carbon is released back into the atmosphere by animals when they breathe and when they decay.

The food chain

A description of how plants and animals feed has been given in the story in Unit 5. We can picture the

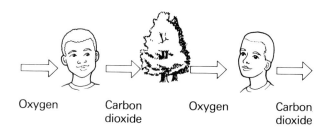

Oxygen   Carbon   Oxygen   Carbon
              dioxide   dioxide

Figure 6 The oxygen cycle

food chain as a pyramid (Figure 7). Each level in the pyramid feeds on the level (or levels) below but, as the pyramid rises, the number of individuals decreases. This is because a greater amount of energy is required to keep the higher levels alive. As individuals in each layer die, they return to the soil to be decomposed, providing food at the base of the pyramid.

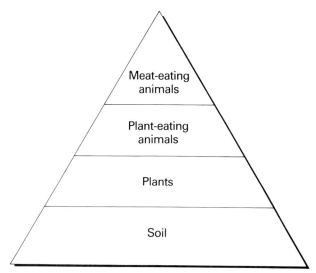

Figure 7 Food chain pyramid

But today people are using up huge quantities of materials that will take thousands of years to replace. Forests untouched for centuries are being cut down for timber, matches, paper and firewood. Oil and coal, products of thousands of years of growth and storage, are being burnt and lost for ever. Mankind is threatening the ecology of the earth by interfering with these delicately balanced recycling systems.

By destroying the main forests, for instance, we lose the trees, and all that the forests contain of plants and wildlife. We are also reducing the ability of the earth to recycle the important air supplies and to produce oxygen.

If a spacecraft runs out of the essential life-supporting supplies or if, in future space missions, the recycling system breaks down, the astronauts will die. If we do the same on spaceship earth, our fate will be the same.

3 Human population

Figures for population growth (Figure 8) are often confusing because it is hard to imagine very large numbers. One way of simplifying the numbers is to use what experts call the *doubling* rate, i.e. the number of years that it takes to double the population of any area. The present doubling rate of the earth's population is about 35 years. In other words in 35 years' time, it is estimated that there will

be twice as many people on earth as there are today. Another way of putting the figures into perspective is to say that at the time when Jesus of Nazareth was alive there were about 250 million people on earth, the same as todays' EEC population. We are *adding* that number every 3–3½ years. How do you feel about facts like these?

An increase in the human population on that scale is bound to have an effect on the rest of the earth's ecosystem. People wonder whether the earth can go on supporting this ever-increasing population. If the millions of starving people are to be saved, then the whole world will need even more food, and a better distribution of that food.

More food means more land; more people need

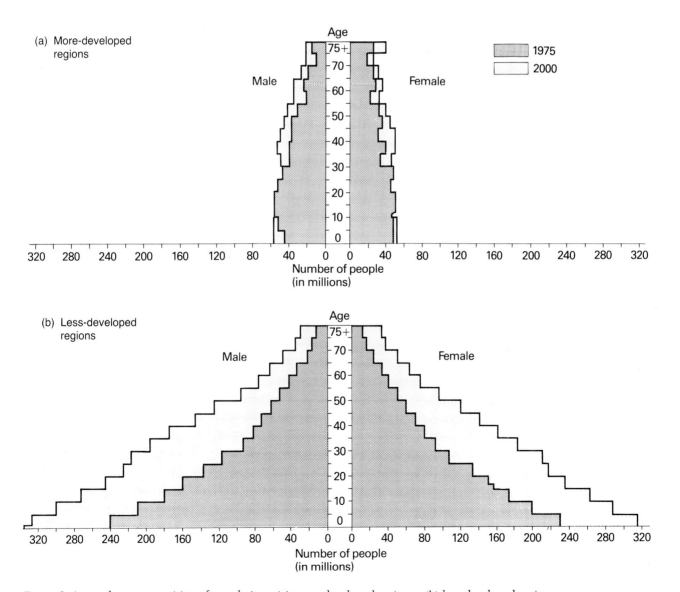

Figure 8 Age and sex composition of population: (a) more-developed regions; (b) less-developed regions

more land for houses. To cope with this growing human population, natural environments which have developed over millions of years are being destroyed.

Strangely enough, it is in the poorest countries where the population is increasing fastest. As Europe gradually became richer, so its population started to decline.

The science of ecology tries to point to what is happening. Other sciences can also help in increasing the yield which farmers get from their crops, but in the end it is people's attitudes and beliefs which will make the most difference.

Discussion

1   It takes years to 'kill' a river or lake, but decades to clean it up. Do you think that human beings know what they are doing? Will we arrive at a silent spring, where no birds sing, or a river where there are no plants or fish? Do people really care? Or is it more important to make money?
2   Are people who live in an unspoilt countryside more civilised or happier than people who live in cities?
3   What kind of beliefs and values are necessary for survival on spaceship earth? What kind of beliefs and values might mean its destruction?

You may like to choose an example to do some role play in class, like the children in the camping story.

In your role play, try to represent the different possible groups (such as industry versus conservation, commercial farming versus organic farming, power stations and city lights versus less lights, and petrol cars versus pedestrians) who might be involved, and let each group argue its case.

4 Responsibility for the environment

Belief and science

Modern science grew up in a society which took certain religious beliefs for granted.

1   God made the world for a purpose.
2   When God made the world, it was good, but it went wrong.
3   Human beings are God's representatives in the world.

These beliefs are found to be especially strong in three of the great religions of the world: Judaism, Christianity and Islam.

The following verses from Genesis 1 illustrate this point.

Genesis 1, 27–31

27   God created *man* in the *image* of himself, in the image of God he created him, male and female he created them.
28   God blessed them, saying to them, 'Be fruitful, multiply, fill the earth and conquer it. Be masters of the fish of the sea, the birds of heaven and all living animals on the earth.'
29   God said, 'See, I give you all the seed-bearing plants that are upon the whole earth, and all the trees with seed-bearing fruit; this shall be your food.
30   To all wild beasts, all birds of heaven and all living reptiles on the earth I give all the foliage of plants for food.' And so it was.
31   God saw all he had made and indeed it was very good. Evening came and morning came: the sixth day.

In Genesis 1, 27, the word man is the Hebrew **'adam** meaning mankind, human race.

In the second creation story in Genesis 2, 7, the same word is used: 'Yahweh (God) fashioned man ('adam) from the dust of the soil ('adamah).' Note the play on words here.

In Genesis 1, 28, the phrase '. . . conquer it. Be masters . . .' means that mankind is to rule over nature as part of his likeness to God, but this does not mean that people should do what they like. They are God's stewards.

Judaism, Christianity and Islam encourage a study of the natural world. Science is not a threat to such beliefs but a natural development from them. Scientists who are religious believers find that these beliefs inspire them in their scientific work and give it more meaning. Indeed, some historians of science argue that, without such belief, modern science could not have developed.

Discuss this; do you think that in order to do science at all a person needs to have certain beliefs? If you say yes, make a list of the beliefs which you think a scientist needs. If you say no, give reasons why you do not agree with the following statement: 'People choose to do things on the basis of what they believe deep down, even if they cannot put it into words'.

Is science to blame?

Some people have said that science, especially in Western countries, is partly to blame for the careless way in which human beings have treated the earth. They point out that, by his intelligence, man has the ability to use the world for his own needs and to exploit all its resources. Does this mean that we have a free hand to do what we like?

When we look at the way that science is being used today, it has not always been for the benefit of the world. The use of science in technology has given our modern world huge benefits, but it has also led to many of our environmental problems.

Some people think that science has made us arrogant. Power over nature has gone to our heads, and this is what causes the problems which we are now facing.

- What do you think of that suggestion?
- Do you think this scientific knowledge has given people the excuse they needed to treat the world as they like? Would we have still treated the world carelessly even if we had not been given that excuse?
- Why should we care for the earth? Is it because human beings will threaten their own survival unless we show more care? Or is that a rather selfish view?
- Do we have a *duty* to care for the earth?

Is religion to blame?

Some people have said that religion, and especially Christianity, is partly to blame for the careless way in which human beings have treated the earth. They point out that in the creation stories in the Bible, man is told to *dominate* the Earth and to *subdue* it. That seems as though he is being given a free hand to do what he likes.

It is sometimes suggested, then, that the religious idea that man is in control of the world and his new-found power in science and technology both combined to cause the problems which we are now facing.

- What do you think of that suggestion?
- Do you think this religious belief has given people the excuse they needed to treat the world as they like? Would we have still treated the world

carelessly even if we had not been given that excuse?

- Do you think that is what the stories in Genesis mean by Adam having dominion over the earth?

Does religion have a say?

A steward is somebody who has been given the job of looking after something. The Jewish, Christian and Islamic religions have always taught that mankind is the steward of God's earth. That is what they mean by being in control. God has entrusted us, they say, to look after the earth and to care for it in such a way that God's creation is not spoilt. We are therefore called to be good stewards of God's earth. First of all, this means that we have to realise that the earth is not *ours* to treat in any way that we want to, regardless of what God intended.

The world and everything in it, including the way that people live, the things that people experience and the events of each person's history, can be seen from two angles: a scientific one and a religious one. It can be seen as facts, and it can be seen as having religious significance. Religious people claim that, just as there is a physical ecosystem, there is also a kind of spiritual ecosystem. This is related to and influences the physical ecosystem.

The way that people conduct their social, family and personal life is important and has consequences, meaning and results. There is therefore a call to balance the material and the spiritual sides of our lives.

Discussion

1 Religion is one area of human activity which concerns people's relationship with God. Can you think of how religious beliefs can change the direction of people's lives? Give examples of people in history whose way of life has been inspired by religion.

2 Find out about people living in the twentieth century, such as Martin Luther King or Mother Teresa, whose actions inspire us to live their kind of life.

3 The world itself and human society seem to need the spiritual inspiration of such people to keep civilisation alive and spiritually alert and prosperous. Draw up a list of beliefs and values, and balance these against material needs.

_7_
Looking at the stars

A view from planet earth

IT was the last night of camp. The fire was out and the campers had managed to find a spot away from the glare of the camp lights. Gradually, their eyes became accustomed to the utter darkness of the night. It was a darkness that some of them had never experienced before. They could not even see the outlines of their hands when stretched at arm's length. The moon had not yet risen, so all that they could see were the stars, countless numbers of them piercing the darkness of the sky.

'That's the view from our home on planet earth,' remarked Mr Payne, as everybody settled down and began to take in the splendour of what they were seeing. 'I don't expect many of you will have seen the night sky like that before. The city lights block out most of the stars, as, of course, the sun does completely during the day.'

'Are they shining all the time, then?' asked someone from the darkness.

'That's right,' he replied, 'but our sun, which we now know is a star and not a particularly special one compared with others that you can see, is so close to us that it seems very bright indeed.'

'So all the other stars are suns also, perhaps having their own planets like earth?' asked one of the others.

'I don't doubt that they have, but they are so far away that we cannot see their planets.'

'Do you think we'll ever be able to visit them?' asked one of the boys.

'I don't know. The difficulties will be enormous. They are so far away. If we wanted to journey to the next nearest star and we could travel at the speed of light, it would take about 4 years,' said Mr Payne.

'That's not long,' said one of the girls.

'No, it isn't, but at the moment our fastest rockets can only reach about one-hundredth of 1% of that speed. So it would take about 40 000 years.'

Patterns in the sky and galaxies in space

There was a long pause as everyone tried to take in these figures. At last Mr Payne said: 'Let's have a look at some of the patterns we can see in the sky. Of course, they're not real patterns. They just look that way to us on earth. They are called constellations.'

'I can't see any patterns,' muttered a boy at the back of the group.

'You probably won't until I point them out to you. Our ancestors, thousands of years ago, decided on the patterns they saw and we have used them ever since. Human beings like to see things in patterns.'

So Mr Payne began to show the group around the sky. He began with the Plough and Orion and then continued with some of the other groups that were not quite so obvious.

'Do you see that misty arch across the sky?' he asked. 'Does anyone know what that is?'

'It's the Milky Way, isn't it?' said Mr Jones.

'Yes. Not all that long ago astronomers thought it was the edge of space, but things have changed a lot since then. Our knowledge of the universe has grown beyond expectation.'

'Do you think we know it all now, then?' asked Carol, whose voice was recognisable although nobody could see her.

'Well, sometimes you hear people say that,' he replied. 'There is always the temptation to believe that our picture of the world is the final one.'

'But,' said Miss Ridgewell, 'it's one of the great lessons of history that, as soon as people start to believe that, someone comes along with a new discovery that challenges the knowledge we think is certain. Then we have to revise our ideas or learn to think bigger.'

'That happened at the time of Copernicus didn't it?' said someone.

'That's right,' repied Mr Payne, 'and it has happened again this century in many fields of science and especially in astronomy. Somehow it seems that, the more we know, the more we know we don't know, if you can understand what I mean.'

'What was the new discovery which started it all this time?' asked Mr Jones.

'It was the discovery that one fuzzy patch of light which had always been believed to be gas turned out to be a galaxy. We then realised that our sun is just one star in such a group.'

'Are there any other galaxies, then?' asked one of the girls.

'Millions of them,' he replied, 'and they are so far away that the light from them takes millions of years to reach us. It's like looking back in time to what they looked like millions of years ago.'

Moon rise

Just then a large ball of light began to appear from behind the far-away hills.

'The moon,' called out someone, obviously very excited . . . .

'That's right. It is moonrise,' said Mr Payne. They all turned and watched, fascinated by what they were seeing.

'At Christmas time in 1966 some astronauts were orbiting the moon,' said Mr Payne, 'and they saw the earth rise like that above the moon's horizon. They were so thrilled by what they were seeing that they sent back a Christmas message. Can you guess what it was?'

'Happy Christmas from the moon, I suppose,' said one of the boys.

'No, it went like this: "In the beginning God created the heavens and the earth. The earth was without form and void, and darkness was upon the face of the deep; and the Spirit of God was moving over the face of the waters. And God said, 'Let there be light'; and there was light.
And God saw that the light was good; and God separated the light from the darkness.
God called the light Day, and the darkness he called Night
And there was evening and there was morning, one day".'

'That's from the Bible,' said Carol.

'That's right,' Mr Payne replied. 'Do you think that it's an unusual message for astronauts to send to people on earth?'

Things to do

Thinking about the story

1　What do you think Mr Payne meant when he said, 'Somehow it seems that the more we know, the more we know that we don't know . . . .'

2   Why did you think the *Apollo 8* astronauts chose the Christmas
     message which they did? Do you think this a good choice or a stupid
     one? What would you have chosen?

3   The very first man in space was the Russian astronaut Yuri Gagarin.
     When he landed back on earth, he said that he had travelled through
     space but had seen no God up there. How does this comment show
     that he knew very little about religion, however clever he was in his
     knowledge of science and technology? How would a religious
     believer have answered him?

Creative responses

Do you find the vastness of the universe strange, mysterious, frightening,
wonderful, exciting, unthinkable or what? Write a poem about this, or
draw a picture, or express your reactions in some other way.

Topic work: The universe and us

1 A greatly expanded universe

Towards the end of the nineteenth century, many people believed that all the basic problems of science had been solved. There was still plenty of work to be done filling in the details of our knowledge but, given time, our understanding of the world would be complete. The universe was thought of as a great machine built on the small solid particles called atoms which behaved according to certain 'laws' such as the law of gravity. Furthermore, Darwin seemed to have solved the problem of how life had evolved on earth.

There were indeed some new discoveries, particularly about atoms and energy which did not quite fit in with existing ideas. People thought that even these were only technical difficulties which would one day be solved without seriously changing the accepted picture of the world.

However, in the twentieth century, scientists have had to change their picture of the world because of some very important discoveries especially in astronomy and physics.

We are going to look at just one or two of those discoveries.

Figure 1 The nebula (cloud) in Andromeda

The milky way

Go out on a dark and cloudless night and look up into the sky and you will see a misty arch stretching across the heavens. We call this the milky way, but what is it?

If you had asked that question at the beginning of this century, you would have been told that modern telescopes show it is made up of countless individual stars. Nobody was sure whether there were other objects in space beyond the milky way, but it was generally thought to mark the edge of the universe – a girdle of stars enclosing all that existed in space. The diameter of this girdle of stars was estimated at 6600 light years. (1 light year is the distance travelled in 1 year at 300 000 kilometres per second which is the speed of light. 1 light year therefore equals 10 000 000 000 000 kilometres. The nearest star (apart from our sun) is 4⅓ light years away from earth.) So the universe was thought to be large but not beyond the bounds of imagination.

Galaxies

Astronomers now realise how hopelessly wrong they had been. In the constellation of Andromeda there is a cloudy patch which can just be seen with the naked eye. It had previously been thought to be a cloud of luminous gas, inside the milky way. Astronomers call it M31 (Figure 1).

In 1920 the American astronomer Edwin Hubble (Figure 2) photographed M31 with the 100-inch telescope at Mount Wilson, Los Angeles. He then looked at the photographs with the help of a magnifying glass. To his amazement he found that the cloud was made up of tiny spots of light. Each spot of light was a star. There were millions of stars, all grouped together in a great spiral, like an island in space *beyond* the milky way. Soon other island galaxies were discovered, hundreds, thousands and then millions of them scattered through space.

Our star, the sun, far from being at the centre of the universe as Copernicus had believed, is just one member of such a galaxy or group. All the stars

Figure 2 Edwin Hubble (right) outside the 100-inch telescope at Mount Wilson

which we see on a clear night are also part of our galaxy, our local group.

This was a most exciting and shattering discovery. It was as though we had grown up in a city, believing that our city was the only one which existed, only to discover there are millions of other cities beyond the boundaries of our own. This discovery has meant stretching our powers of imagination almost beyond anything that we can possibly think of. Our thinking had been too small, minutely small, compared with the vastness of space which is gradually being revealed.

Figure 3 Spinning tops

How can we begin to picture the size of the universe as it is known today? It is an almost impossible task. All we can do is to scale it down to a model.

Imagine that our own galaxy is a spinning top 1 metre in diameter (Figure 3). M31 would then be a similar top about 15 metres away. Now imagine several thousand million tops each about 15–20 metres apart filling a huge ball 150 kilometres in diameter. That will give you a very rough idea of its size. Our earth would then be as small as one of the molecules which form our spinning top.

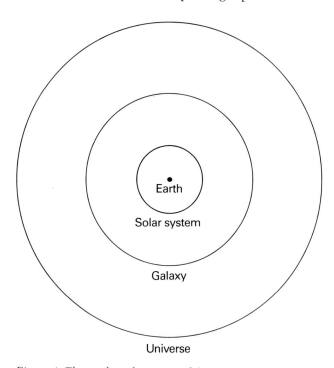

Figure 4 The earth as the centre of the universe

Figure 4 gives the impression that earth is the centre of the universe.

- Is this so?
- Why has it been drawn like this?
- Can you put the same information into a diagram which does not give such a wrong impression?

2 A universe flying apart

Shortly after Hubble had discovered that there are other galaxies in space, he noticed something else which seemed even more surprising. All the galaxies appear to be flying apart. It is as though they are fragments from a great explosion. When an

Figure 5 As the balloon expands, the spots move further apart and further away from the centre of the balloon

explosion happens, pieces of debris are thrown out in all directions. Hubble observed that the galaxies farthest from us in space seem to be going faster. Indeed, all the galaxies (ours included) are speeding away from all the others.

This is difficult to grasp but a very rough idea can be gained from the balloon drawing (Figure 5). Can you think of any other ways of expressing this?

Seeing the universe with the help of a theory
Of course we cannot see this happening in the same way that we see a racing car moving along the road. This 'observation' is based on a theory. It has sometimes been nicknamed the 'hee-haw' theory.

Imagine that your friend is speeding along on a train and the driver sounds the whistle. Your friend on the train hears the whistle as a steady note. If you were waiting at the side of the track you would hear a different sound. As the train approached, you would hear a high pitched 'hee' sound, and as it passed and went into the distance, the sound would drop in pitch to a 'haw' sound.

The reason for this change in pitch was described by the Austrian scientist Christian Doppler. It is called the Doppler effect. Figure 7 shows what is happening. The sound waves coming towards you are bunched up – their frequency increases. But as

(a) On a clear night, many stars can be seen by the naked eye

(b) The image seen through a telescope is 40–100 times larger than that seen with the naked eye

(c) A radio telescope can enable us to see deep into space, beyond any visible range

Figure 6 Extending vision

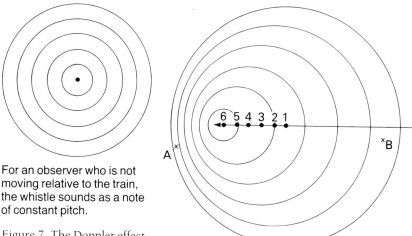

For an observer who is not moving relative to the train, the whistle sounds as a note of constant pitch.

Figure 7 The Doppler effect

For an observer at A, the source of the sound (the whistle) is getting nearer as the train approaches. This has the effect of bunching up the waves. The observer hears a constant note but of higher frequency.

Similarly, for an observer at B, the source of the sound is moving away. The sound waves reach the observer at a lower rate. So the observer hears a note of lower frequency.

the train goes away from you the sound waves reach you less frequently – their frequency decreases. High frequencies give high-pitched sounds; low frequencies give low-pitched sounds.

The first experimental test of Doppler's principle was made in 1845 in Holland. A train carried a group of trumpeters to and fro past some musicians (Figure 8). The musicians were able to hear the difference in the pitch of the notes being played depending on whether the trumpeters were moving towards them or going away.

Light travels in waves just as sound does; so the Doppler effect also works for light waves. The light coming towards you increases in frequency and so becomes more blue. The light going away from you decreases in frequency and becomes more red. When we look at all the galaxies in space, we see that their light is becoming more red. This means, according to the Doppler effect theory, that they are all going away from us.

In the beginning

Most astronomers today agree with what is called the 'big bang' theory (Figure 9). According to this, our universe of space and time began with a big bang over 15 thousands million years ago. Before that, there was nothing at all – no space, no time and no matter.

3½ minutes after the explosion the basic building blocks of the universe were formed (the nuclei of the atoms of hydrogen and helium).

1 million years later the first atoms were formed (these were the atoms of hydrogen and helium when electrons joined with the nuclei).

Millions of years later these atoms began to form into vast clouds, slowly spinning. These clouds were the beginning of galaxies. (Gravity drew the atoms closer together and, just as when an ice skater spins faster her arms are drawn closer to her body, so the young galaxies began to spin faster.) Within these young galaxies, smaller clouds of gas developed. The

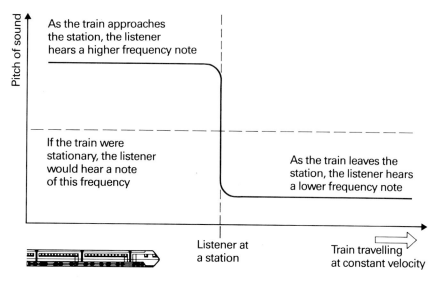

Figure 8 The Doppler effect for a train travelling through a station

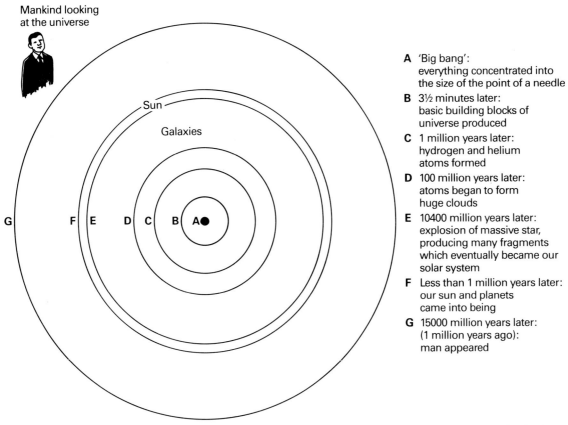

Mankind looking
at the universe

Sun

Galaxies

G   F E   D C   B A●

A 'Big bang':
everything concentrated into
the size of the point of a needle

B 3½ minutes later:
basic building blocks of
universe produced

C 1 million years later:
hydrogen and helium
atoms formed

D 100 million years later:
atoms began to form
huge clouds

E 10 400 million years later:
explosion of massive star,
producing many fragments
which eventually became our
solar system

F Less than 1 million years later:
our sun and planets
came into being

G 15 000 million years later:
(1 million years ago):
man appeared

Figure 9 The 'big bang' theory

effect of gravity on them not only made them spin faster but also made them become very hot indeed, so hot that nuclear reactions occurred which produced the first stars. These then began to explode, sending out helium, carbon, oxygen and other heavy elements. From these elements, other stars and planets were formed.

About 10 400 million years after the 'big bang' a massive star exploded showering space with its star dust.

Less than a million years later (4600 million years ago) a small yellow star and its nine planets came into being.

Eventually, 15 000 million years after the 'big bang', a two-legged creature with remarkable capacities appeared on one of those nine planets.

Many astronomers make clear that they accept the 'big bang' theory at the moment because it seems to fit the facts as we have them now, but they realise that more information may turn up which will make them change their theory.

Some years ago, astronomers held a different theory, called the steady state theory. Instead of a huge explosion which marked the beginning of

creation, the hydrogen atoms which make up most of the universe are said to be continuously created. Strange as it may seem, this theory states that the universe never had a beginning nor will it end. As the universe expands, the space which is left 'empty' is continuously filled with new hydrogen atoms which are then gradually drawn together to form new stars and galaxies. This theory has ceased to be popular.

3 Is your picture of 'God' too small?

Just as we have learnt to develop our concepts in science, so in religion we must learn to develop our understanding about God.

In Unit 5 we looked at some pictures of God which are very inadequate and childish.

Write down how you think that religious people picture God. Compare what you have written with the idea of God given in the rest of this section.

'God exists'
For believers, God is not just something people have

made up. God is not just an idea. God is a Being who is Personal. But God's existence is different in an important respect from the way in which people and objects exist.

When we say that this table exits, this rose exists and you exist, we are talking about things or persons which exist in time and space. In other words, there was a *time* when this table, this rose and you did not exist, and there will be a *time* when these will cease to exist. This kind of existence begins and ends. Similarly, these things and persons all have bodies which occupy *space* and have shape, colour, weight etc. This kind of existence is in one place and not in another.

Here is how a religious person thinks of God. Discuss this view. 'God, however, eternally *is*. God is not an object or person in time and space. God's existence has no beginning and no end; there never was a time when God was not and there never will be a time when God is not. Nor does God have a body.'

'God is indescribable'

This follows from the fact that God is not an object or a person in time and space. We cannot examine God under the microscope or weigh God on a pair of scales.

But people claim to have experience of God. It is as though God sets up spiritual communication lines so to speak so that, if they want to 'listen in' they can come to know something about God. How can they translate this knowledge, which they acquire spiritually, into words, for themselves or others? How can they make sense of it? describe it?

Religious people have discovered at least two ways in which they can begin very inadequately to describe what God is like. One of these ways is by saying what God is not, e.g. that God is indescribable, which means that God is *so much more* than they can find words for.

Find out what the following words mean which are used to describe God: invisible, infinite, incomprehensible, ineffable, unchanging. Can you think of any other? This way of talking about God is often called the *via negativa* (Figure 10), the negative way (*via* meaning way as in 'viaduct').

'God is like a father'

Another way of talking about God is by comparing God with things and people that we do know about, i.e. by using a **metaphor**. Religious people may say, for example, that God is 'He' and not 'It' and that 'He'

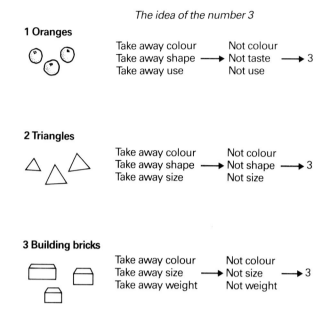

The idea of the number 3

Figure 10 The *via negativa* is a way (via) of knowing about something (in this case, the number 3) by saying what it is not (negativa)

sees, acts and feels like people do. They may say that God is like a ruler in his power, an artist in his creation of things which are beautiful, a father in his caring for the welfare of his children etc.

This is similar to the way that language is often used in science; when we talk of 'waves' to describe how sound and light travel, we are using a metaphor. Sound and light are not the same as water waves, but it is helpful for us to compare them with waves of water.

In religion as in science, people explain what is unfamiliar by speaking of what is familiar. Can you see what may be a danger in this? (Look back at Unit 1.)

When we say 'Our Father', we are using a metaphor; we mean that God is like a father to us all. There is a danger that some may even misunderstand this, because some people have experience of the word 'father' which is not happy.

Here are six comments by children asked to write a sentence about how they see their father.

1 'I'm frightened of my father – my mum is too. He often comes home drunk, and sometimes he has an awful temper.'
2 'My father and I get on ever so well together.'
3 'My father wouldn't know whether I was at home or not. He's always out.'
4 'I haven't got a father.'

5 'I love my father because he's ever so kind and cheerful and helps Mummy to do the housework when she's not well.'

6 'My father's unfair. He locked me in my bedroom last week for telling a lie – and I never.'

Each of these comments comes from the child's unique experience, and this experience will colour the view of 'father' which each child has. What picture of God will such people tend to have if God is talked of as a father?

Sort them out from the following list by matching the six comments 1–6 with the statements (a)–(f) below.

(a) Someone who is non-existent.
(b) Someone who is a friend.
(c) Someone who is unjust.
(d) Someone who is good.
(e) Someone who is unreliable.
(f) Someone who is absent.

Which of these children would most easily be able to understand what religious people mean by calling God 'Father'?

'God is creator'

When people believe that God is creator, they mean that God is in control of the whole system. God is all powerful: God created the world not out of anything! He made everything that is. God is *not* like a builder building a house or a sculptor sculpting a statue, for they make these things out of other material. God created time and space; without God there would be nothing.

When religious people think about the immensity of the universe (Figure 11), it awakens in them a desire to praise God. Here are some examples. The first is from a Hebrew writer.

Psalm 104, 1–5 (The Bible, Revised Standard Version)
1 Bless the Lord, O my soul. O Lord my God, thou art very great; thou art clothed with honour and majesty.
2 Who coverest thyself with light as with a garment, who hast stretched out the heavens like a tent.
3 Who has laid the beams of thy chambers on the waters,
   who makest the clouds thy chariot,
   who ridest on the wings of the wind.
4 Who makest the winds thy messengers, fire and flame thy ministers.

Figure 11

5 Thou didst set the earth on its foundations, so that it should never be shaken.

Next consider an example from Islam.

Qur'an 2, 255
'There is no God but Allah
the Everlasting, the Self-existent
by whom all else exist.
Slumber overtakes Him not, nor sleep.
To Him belongs whatever is in the heavens
and whatever is in the earth . . . .'

Figure 12 Muslims at prayer

Figure 13 A first-century Jew

Here is an example from a Christian writer.

Romans, 8, 38–39 (The Bible, Today's English Version)
38 For I am certain that nothing can separate us from God's love: neither death nor life: neither angels nor other heavenly rulers or powers; neither the present nor the future.
39 Neither the world above nor the world below – there is nothing in all creation that will ever be able to separate us from the love of God which is ours through Christ Jesus our Lord.

The writer was Paul, one of the first Christian saints, and he was writing to Christians in Rome about AD57. Would he have been worried by the findings of modern astronomy and physics? Why or why not?

Finally we have a Hindu example.

'God created the universe, like a musical instrument and filled it with His word as it sounds. God created the eternal Om out of Himself.'

When Hindus pray, they often pronounce the word Om (Figure 14). Om is the sacred syllable which Hindus believe was spoken at the foundation of the universe. They believe that, if they say it slowly and reverently and with lengthy resonance, it will bring great blessing.

Figure 14 The sacred syllable Om

Science raises questions which go beyond science
Scientists find out all they can about the natural world and how it works, but religious people are concerned about the why, the wherefore and the meaning of everything which scientists discover. These are questions of purpose and meaning which science cannot answer but which science often helps people to ask.

One scientist says this about it: 'Modern science insistently forces upon us the questions going beyond science, questions that can only receive a theological answer.'

Here is a comment by another scientist: 'It's not for astronomers to say what caused the big bang . . . . That's not in the rules of the game. All they can say is that the big bang happened, setting into motion our expanding universe.'

Study this and say how a religious person might see this as opening up the way to religion.

The immensity of the universe and the importance of people
When some people think about the kind of things which this unit has been dealing with, they feel that human beings are so small and unimportant (Figure 15).

Does the immensity of the universe mean that people do not matter?

Think about this in the light of what a great scientist and thinker of the seventeenth century, Blaise Pascal, wrote: 'Man is but a reed, the most feeble thing in nature. The entire universe need not arm itself to crush him; a vapour, a drop of water, is sufficient to kill him . . . . Yet man is a thinking reed. If the universe were to crush him, man would still be nobler than that which killed him, because he knows that he dies, and the advantages the universe has over him; of this the universe knows nothing.'

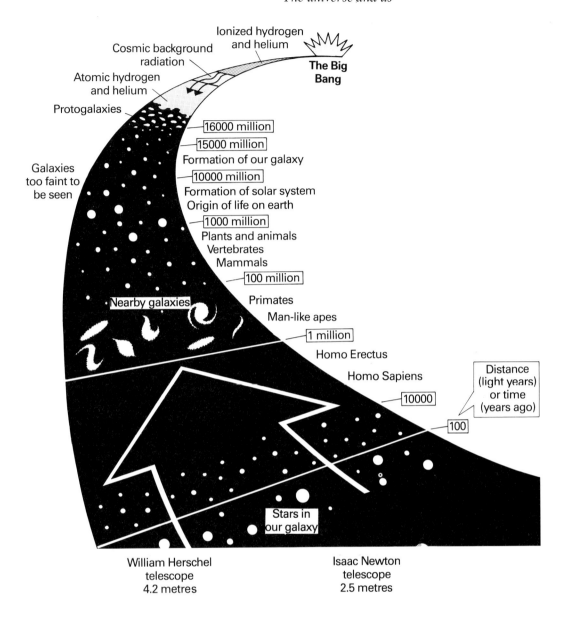

Figure 15 Looking back in time

- Can you put in simple language what Pascal is saying here?
- Do you agree with him?

- Do you think that people are something more than just the chemicals of which their bodies are made? Give as many reasons for your answer as you can.

8
Home from home

The last day of camp

*T*HE last day of camp arrived. As the campers packed their bags, they felt quite nostalgic.

'It's really been like home here,' volunteered Carol. 'I don't want to go back.'

'Yes,' said Winston, 'it's been fun, but it's also been sort of peaceful. There's been time to live, instead of all the rushing about from one thing to another.'

'And the discussions have been good,' said Dawn. 'Pity that school can't be run more like this. I mean – being able to talk about real questions and trying to find your own point of view.'

'Even if you are often the odd one out!' chuckled Wayne.

'I guess we're all unique,' commented Miss Ridgewell. 'No-one else has had quite my experience or yours, Wayne, or Carol's, or anyone else's.'

'It's just that some people talk so much more than others,' said Dawn, looking at Wayne, and there was a ripple of laughter all round.

'But Wayne has said some interesting things,' observed Mr Payne. 'I'm glad he came.'

'Even though you often don't agree with him?' asked Leila.

'Oh yes,' replied Mr Payne, 'that doesn't matter so long as there is real communication between people. And Wayne has raised some big questions.'

Wayne beamed and said how much he had enjoyed it. 'It's been like leaving home to go home,' he said.

'Yes, it's funny,' said someone else. 'What makes a home a home?'

'Well, it's feeling at home,' said one.

'It's being accepted for what you are,' said another.

'Yes,' said another voice, 'here at camp we've made it home because we all got on very well together.'

'I don't think that's true about my home,' said Wayne. 'We're all at sixes and sevens – like a jigsaw puzzle where all the bits don't fit.'

There was a moment's silence after that. No-one knew quite what to say, although they knew what he meant.

School assembly

Just then Mr Jones turned up to make an announcement: 'The Head has asked us to run an assembly next week. He said that we could share with the others any fresh ideas we had as a result of coming to camp. So I suggest that you start thinking about it on the way back. Then tomorrow afternoon we'll have a meeting to plan it.'

Something of a silence greeted these words, followed by quite a muttering.

'Must we?' asked someone. 'Assemblies are so boring.'

'Yes, and a lot of people don't believe in this God-stuff,' said someone else, 'I for one.'

'It'll just spoil it all – having to talk about it,' said another. 'People don't like assemblies. They feel they're being got at with religion.'

'Well,' said Mr Jones when they had all finished their grumbling, 'you're just the people to put on a really first-class assembly because you are aware of the problems.'

'You mean that, if it's boring, it's our fault?' asked Carol.

'Well, yes,' said Mr Payne, 'and instead of it spoiling the memories of the week it could be a good excuse to enjoy them again.'

'The trouble is,' said Wayne, 'that, when we're in assembly, everyone seems to take it for granted that you're religious when you're not. You want to feel free.'

'Yes, I agree,' said Mr Payne. 'I think our assemblies should be reformed to make it clear that we don't expect everyone to agree. What we do want is for everyone to think about the questions for themselves. Would you find that kind of assembly acceptable?'

Vigorous nods of approval all round showed Mr Payne that he had hit the nail on the head.

'So your assembly will be a model,' he said. 'You'll show the rest of the school how they should really be run.'

'No prayers or hymns,' said someone.

'Oh, I like hymns . . .,' said someone else.

'Well, you can't ask just the religious ones to sing a solo, can you?' piped up another voice.

'But', said Winston, 'what's wrong with prayers and hymns if you introduce them right?'

'How do you mean?' asked Mr Jones.

'Instead of saying, "Now we'll sing hymn number 207 *Summer suns are glowing . . .*",' replied Winston imitating exactly the Head's voice so that they all laughed, 'why not add, "Think about it while you're singing and why you agree or don't agree with it"?'

'Yes, and a prayer could be something you just listen to or something which you pray,' said Carol.

'And the options must be made clear to everyone, mustn't they?' summed up Mr Payne.

After this talk about assemblies they all felt much better about it, and they actually began to share ideas for it as they travelled back. One person suggested that they do a storm and build it up with slides and sound effects. They could read a poem that Dawn had written! They

could act the play about the Earth King and attack him for allowing bad things to happen.

Someone else wanted to black out the hall and to simulate a journey back in time to the beginning of the world, leaving people with the questions: 'Was there anything before that?' 'Why is it here at all?'

Another group wanted the assembly to raise the question: 'Should we feel at home on planet earth?'

Eventually they worked it out, and the following week the assembly was held. When it came to an end, everyone remained completely silent. Perhaps that was a bigger compliment than they could have hoped for.

Things to do

Thinking about the story

1 How do you see the place of assemblies in school? Is their purpose to help people to think about important questions for themselves?

2 Should religion be part of assemblies, at least sometimes? How can readings from scripture, prayers and hymns be introduced so as to make clear that people are free to agree or disagree with what they say? (Winston gave an example in the story.)

Creative responses

1 Complete the story by compiling the assembly on science and religion which the youngsters did so well.

2 'I've learnt more about myself,' volunteered one of the girls.
'I don't know what you mean,' said someone else.

'Well, although there've been people around all the time, I've had lots of private thoughts. I've never really looked at the sky or the countryside in the way I've done at camp – it sort of makes you stop and think'.

'Yes,' said Miss Ridgewell, 'I feel the same. Sometimes perhaps we just need to be alone to think.'

Continue this conversation in class, giving your own thoughts after reading about the camp and all these units of work!

Topic work: The relationship between science and religion

When you have read the following section, copy and complete the chart below.

Subject	Reason	Reply
The storm	7	(e)
The accident		

Figure 1

1 How are science and religion related?

This book has been about the relationship between science and religion. People have very different views about this and you will need to go on thinking about this for yourself.

It is important, however, that we keep cool and think straight about it. When people misunderstand other people, it can cause serious problems including prejudice, hatred and war. The same thing can happen when people misunderstand *ideas*.

When people think about science and religion, many disagreements can crop up. For example, we can make a list of reasons why some people say they do not believe in God.

1 'Religion is the enemy of science – look what happened to Copernicus and Galileo.'
2 'Science will be able to explain everything – we don't need religion.'
3 'We are here because of evolution and not because of God.'
4 'The enormous scale of everything just shows that there isn't a God.'
5 'The Bible is rubbish because it says that God made the world in 6 days.'
6 'What matters is that people should use their intelligence and not rely on an imaginary God.'
7 'There's such a nasty side to the natural world that there can't be a God.'

Each of these seven reasons comes into the units in this book. Each unit has given information and questions which show the meaning of these seven reasons, and how each one is based on a misunderstanding about religion.

- Match up the seven units with the seven reasons 1–7 writing the number in the chart of Units 1–7 given in Figure 1.
- Explain why each reason can lead to controversy.

Now read the possible replies by a religious person given below.

(a) God has chosen to work through people.
(b) God creates by using evolution.
(c) God is outside time and space altogether.
(d) Miracles happen for a particular purpose.
(e) If suffering suggests that there is not a God, happiness, joy and ecstasy suggest much more powerfully that there is a God.
(f) Religious people have sometimes opposed science but for reasons which were not religious.

(g) The Bible uses picture language to express what would otherwise be very difficult to understand.

- Match the seven replies (a)–(g) with the seven reasons 1–7 above.
- Insert the letters (a)–(g) in the chart in Figure 1.
- Explain how each of the religious replies (a)–(g) tries to answer the controversies 1–7 about science.
- Show how a 'religious' person can accept all the findings of science and the teachings of religion.

What is important?
Do you see the following as important?

Figure 2

Wonder can be felt by a person watching a beautiful sunset or a baby thrush or the intricate detail of a crystal under a microscope (Figure 2).

People can and do show kindness, generosity, unselfishness and love (Figure 3).

Figure 3 A helping hand

Figure 4 Size is not everything

Size is not everything – a person is smaller than an elephant (Figure 4) or a mountain or a galaxy and yet the person can know about these things.

Consciousness (Figure 5): it is strange that we can think about ourselves and ask, 'Who am I?'

People are curious and love to ask questions not just 'How does it work?' but 'What's the point of it all?', 'What's the meaning and purpose of life'? and 'Should we feel at home on planet earth?'

Many people do not rate these sorts of experience very highly. They think that they are not reliable signposts to the nature of life. But many people,

Figure 5 Who am I?

including scientists, do see these as very important. It is for each person to explore these things for themselves.

> Einstein said, 'Religion without science is lame; science without religion is blind.' What do you think that Einstein meant by this, and do you agree with it?

Figure 6 Curiosity

Look back over the story at the beginning of this unit and recall how the campers used their imagination when preparing the assembly. Their camp experience was so important to them. They wanted the school to share in that camp experience; they wanted to make it real for those who had not been to the camp.

To do this, they acted out, as a 'play', the things which they had experienced themselves.

We often make things more real by using our imagination. You yourself may never have actually been to a school camp. In this book we have been reading a story about a school camp. Stories are a way of using our imagination to make things more real and interesting.

We shall now see how in science we use our imagination to make sense (or theories) about the facts, and how we use our imagination in writing, reading and thinking about religion.

The following sections may only be suitable for more advanced pupils.

2 Imagination and science

Common-sense ideas of time and space

In Unit 7 we learnt about our modern picture of the world. We have seen how our understanding of the size of the universe and the structure of atoms has changed dramatically this century. We have discovered a world which stretches the human imagination to its limits. In this section we are going to try to stretch our minds even further, for the world of space and time turns out to be very strange indeed. In a moment you will need to fasten your seat belts to retrace one of the most exciting journeys into human knowledge that has ever been made.

First of all, before we set off on our journey of imagination, we need to make some preparations. Nothing seems more obvious than our common-sense ideas of space and time. Let us take the idea of space first. Everybody knows what is meant by a millimetre, a centimetre and a kilometre. We can easily measure the distance between London and Glasgow or for that matter between our sun and the next-nearest star. Perhaps that is not quite so easy, but at least we know what it means. If we were to take a journey out into space, we would expect it to go on and on and the distance between stars would be the same whoever is measuring it.

The same can be said of time. We measure different lengths of time on our watches. We can look on a timetable to see how long it takes to travel from London to Glasgow by train. If you and your friend were to time the same journey and find that you did not agree, then it would be obvious that one of you had made a mistake. Time appears to move at a steady pace and to be the same for everybody and in all places.

This common-sense view of space and time is more or less correct for everyday purposes but *not* if we want to be accurate in our knowledge of the universe.

Einstein on time and space

Just before the turn of the century a young man, only 16 years old, started to ask some very simple questions about our common-sense ideas. By 1905 he had reached certain conclusions which began a

Figure 7 Albert Einstein

major revolution in science – greater even than the change which took place at the time of Copernicus. His name was Albert Einstein (Figure 7), and his theory is called relativity.

Our attempt to understand Einstein's theory of relativity will avoid all the difficult mathematics which he used. Instead we can try to imagine what it tells us about the world, for Einstein said, 'Imagination is more important than knowledge.'

We can start our journey of exploration with a simple observation. If you have ever travelled on a train or in an aeroplane, you will know that, when it is going at a steady speed in a straight line, it is hard to tell that you are moving. The laws of nature are just the same whether you are standing still on the earth or moving at a steady speed in the aeroplane. If you walk to the front of the aeroplane, you do not have to walk faster because the aeroplane is moving forwards. In other words, if the aeroplane is moving at 600 kilometres per hour and you want to walk forwards at 2 kilometres per hour, you do not have to run 602 kilometres per hour to get to the front. But suppose that somebody was not on the aeroplane but was watching you from the ground and wanted to measure your speed as you approached him. Would he say that you were travelling at 2 kilometres per hour, 600 kilometres per hour, or 602 kilometres per hour?

While you are thinking about the answer, we shall mention another idea which occurred to Einstein.

All the experiments which have ever been made on the speed of light have confirmed that light travels at 300 000 kilometres per second in a vacuum. Einstein wondered about this and asked himself the question: would the speed of a beam of light change if someone was moving towards it? Would the light reach him quicker or would it still be 300 000 kilometres per second? In the aeroplane example given above, you have probably realised that you add the speeds together to get the answer. To an observer on the ground you are moving at 602 kilometres per hour. Should we do the same for a *pulse* of light as we move towards it? The obvious answer is that you *would* need to add it on and that is how it had always been understood. It turns out, however, that this is not true. Light is different. It always travels at 300 000 kilometres per second whatever speed you are travelling towards or away from it. This is very strange, and it leads to even stranger things. To begin with, this means that in our previous example of the aeroplane the correct answer is not 602 km per hour but 601.999 999 999 999 999 999 4 km per hour. The reason is simply that our normal ideas of speed and distance are only approximately true. At slow speeds the difference is not very great but as we approach the speed of light, the differences become very great indeed. What happens is that our everyday ideas of time and distance have to change.

What is speed? It is distance (e.g. 600 kilometres) travelled in one unit of time (e.g. 1 hour). If our understanding of speed is no longer accurate, that makes a difference to our ideas of time and space.

Relativity and time

Imagine that it were possible for an astronaut to travel away from earth approaching the speed of light. As he reached higher and higher speeds, his clock in the spacecraft would begin to slow down relative to an observer on the earth. In fact he would slow down too – his heartbeat, his breathing and his thinking would all be slower relative to the observer on the earth. He would even age less than his twin brother on earth. If he reached the speed of light, he would not age at all! But here comes the real crunch – he would not be aware of any change. For him, life would be going on quite normally.

Relativity and space

Things are just as strange when we think of space. Our astronaut sets off from earth and we watch him as he goes. He also sees the earth speeding away from him at the same rate. Because he is travelling close to the speed of light, his time is running slower. We notice that it takes, say, 30 years to reach a nearby star, but according to his time it has only taken 1 year. In other words, on his calculations, he has only travelled one-thirtieth of the distance that we have seen him travel. For him, space is smaller than it is for us!

Relativity and gravity

The strange effects of relativity on our everyday ideas lead to the conclusion that space and time are not separate but are linked together. Einstein went on to think about the effects of gravity on space and time. Close to a massive body such as the sun the shortest distance between two points is not a straight line but a curve. It could be that the gravitational attraction of all matter in the universe curves the universe itself so that it does not go on for ever and yet at the same time there is nothing beyond it. It has no edge.

To give you some idea of how things like that can happen, imagine a beetle on a large surface. The beetle decides that it is going to crawl in a straight line until it reaches the edge of the surface. On and on it goes until one day it finds itself – back exactly where it started. How could that happen? Quite easily, if the surface turned out to be a football. The beetle is completely baffled. It can only think in two dimensions, but the football is in three dimensions. Our brains are so designed that we think in three dimensions. We describe things as being up or down, left or right, or backwards or forwards. You cannot think in four dimensions and so there is no need to try. It does seem as though the universe is created in four dimensions, perhaps even more, and that this fourth dimension, 'space–time', might well curve back on itself so that, if our astronaut set off to find the edge of space, he would one day return home to planet earth without arriving at his intended destination.

A spiritual dimension: a new name for an old idea

The question can indeed be asked about whether there is another dimension, the dimension of the spiritual. All the great religions of the world have over the centuries given evidence of such a dimension and affirm the possibility of experiencing that which is beyond time and space altogether. Many individual people have had an awareness of a Presence which is at the same time quite separate from themselves and yet which is closer to them than their own breathing. The Presence is felt to be both transcendent and immanent. **Transcendent** means that this Presence is outside us, greater than anything in time and space: something our minds are not able to understand even though we can know that it is there. **Immanent** means that this spiritual reality is felt to be also within us: the spirit is at work in and through everything in the world.

3 From the very large to the very small

When we turn from the very large objects in space to the very small particles of which all matter is composed, we also find that our modern ideas are very different from those in the nineteenth century. About 100 years ago it was generally believed that everything in the world was made up of very small *solid* particles of matter called atoms. Atoms were known to combine together to make larger particles called molecules. As there are many different kinds of atom, the various ways in which they combine together produce a whole variety of different molecules. This is the reason that we find such a variety of substances on earth, e.g. wood, minerals, tissues etc. For instance, wood is made of atoms of oxygen, hydrogen and carbon.

The idea that everything is basically made of tiny *solid* particles was first suggested by the ancient Greeks and has a certain ring of common sense about it. After all, when we go to the beach, we see that it is made up of tiny particles of sand, and it is natural to think that, if these particles were broken down any smaller, we would arrive at a particle so small that it cannot be broken down any further.

By the beginning of this century about 80 different atoms were known to exist. (The number is now

well over 100.) They are very small, about a ten-millionth of a millimetre across, and they *are* the smallest form which any of the basic substances can have. For instance the smallest piece of lead is an atom of lead. During this century, however, we have discovered that they are *not solid* and are not the smallest building blocks of matter as they were thought to be. If we think of our example of lead again, an atom of lead turns out to be made up of even smaller particles although they no longer behave like lead any more.

Waves or particles
The story of the atom is given in Figure 8 in the form of pictures. The pictures will help you to understand how our knowledge of atoms has changed but there is one very large snag. As we get nearer to our present-day understanding, it becomes increasingly difficult to draw what we think atoms are. This is not because the artist is no good. It is because the physicists, who investigate these small particles, have discovered that it is much easier to say what they *do* rather than what they *are*. In fact it has seemed to many of them that the questions 'What are they?' and 'What do they look like?' are the wrong questions. We can *only* say what they behave like.

If you find that rather worrying, you will be pleased to know that you are not alone. For more than 20 years, two of this century's greatest physicists argued very passionately about whether science can say what the world is like any more. We shall try to see, very briefly, how this argument came about.

We have said something about the nature of light in our previous section on relativity. We start with a problem about light in this story. In the nineteenth century every scientist believed that he knew what

Figure 8 The story of the atom

 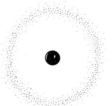 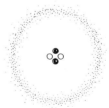

(a) In the nineteenth century the atom was thought to be solid and the smallest particle, from which everything was made

(b) In 1897 the electron was discovered. This showed that atoms were not solid, and in 1911 it was suggested that the atom looked like a mini solar system

(c) In 1913 it was found that electrons do not (like planets) keep to their orbits but are in fact a 'cloud'

(d) In 1932 it was discovered that the nucleus is also not solid but is made up of protons and neutrons

(e) Now it is known that even protons and neutrons are not solid but are made up of smaller particles called quarks

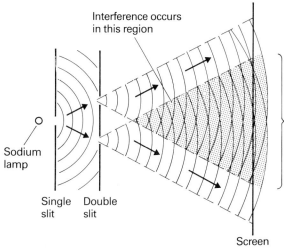

Interference occurs
in this region

Sodium
lamp

Single   Double
slit   slit

Screen

For interference fringes to be seen, the
light coming from two sources must be
coherent — it must have the same frequency
and wavelength and it must be exactly in
phase.

Light and dark bands
called interference fringes
are seen on this
part of the screen

Young found that he could not achieve this
with two separate lamps, so he used as his
light sources two slits illuminated by a
single lamp. For the experiment to work,
the slits must be very narrow and very close
together (about 0.5mm apart).

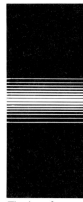

The interference
pattern obtained

Figure 9 Young's double-slit experiment to show the interference of light

light was. Light was thought to be made up of waves, rather like the waves which we see when a stone is dropped into a pond. This belief was confirmed by experiments such as that in Figure 9. In the seventeenth century, Isaac Newton had said that light was made up of particles and for a while his view was accepted. But this experiment seemed to show that he must have been wrong. Surely light must be waves and not particles?

Then came an anomaly – every experiment *except one* showed that light seemed to be made up of waves. Not for the first time scientists found themselves with a problem. The question was: 'Should they worry about this one anomaly?' The scientist who argued that they should worry was Albert Einstein.

A solution to the wave–particle anomaly was suggested by the Danish physicist, Niels Bohr. 'We cannot', he said, 'say what light *is* or electrons *are*, we can only explain what they do in this experiment or that experiment. And what *we* do in the experiment makes a difference to how they behave.'

It was this solution which caused the long argument that we referred to earlier. Bohr believed that science can no longer say what nature is really like but only how it behaves. Einstein could never accept this. He believed that despite all the problems we would one day be able to explain what it is really like. This has important consequences for the actual conduct of scientific research.

Bohr regarded the quantum theory as a complete explanation, believing that you can ask questions about the behaviour of electrons but all reference to

an underlying reality is excluded and meaningless.

Einstein, however, regarded quantum theory as a partial theory, a stepping stone to further questioning and a more complete theory. He believed that 'something deeply hidden had to be behind things.'

4 Imagination and religion

Our imagination can be used in religion, as it is used by scientists, and so help us in our understanding of God. Just as we asked the questions, 'What are electrons?' and 'What are electrons like?' so we can ask, 'Who is God?' and 'What is God like?' If science cannot yet say what electrons *are* and can only describe in metaphors how they behave, perhaps we ought to do the same about God. What we should learn to describe in metaphors is how God is believed to behave in the world and especially how God does behave towards us as persons.

The language of religion uses such metaphors and the imagination to express ideas for which ordinary words are inadequate. Think, for example, of a person whom you know well and admire. Try to describe that person's character to someone who does not know him or her. Is it easier to say what such persons are in themselves, or to describe what they do and how they behave towards you and other people?

We do this all the time in our everyday life. We do it especially in literature, in poetry and in works of the imagination. For example, if a man went around

Figure 10 John Gielgud as Macbeth

imagining that he was a Roman soldier, Henry VIII or Macbeth, we would say that he was mad but, if it were a play in which he was acting a part (Figure 10), we would say that he was helping the audience to feel what it would have been like to be a Roman soldier, Henry VIII or Macbeth. He would be trying to make those worlds real for us today. Whether or not this helps him and the audience *really* to see what it was like to be a Roman soldier, Henry VIII or Macbeth depends on such things as how good the play is and how good an historian the author is.

Actually the only way that we can know about the Roman world, the Tudor world or the world of mediaeval Scotland is by using our imagination to 'clothe' the plain information which historians give us. We all know the difference between someone repeating 'the Romans built Hadrian's Wall' as a dead 'fact' of history and someone saying the same thing delightedly, with enthusiasm for the great engineering skill which the Romans displayed in building Hadrian's Wall.

Revelation and imagination
The idea of what God is really like in Himself could not have been guessed. According to the Christian religion, this inner life of God has been told to people through a process called **revelation**.

Revelation is a process known in most religions. It means that God tells us something that could not otherwise be known. In Islam the author of revelation is the personal God whose will is expressed as commandments. The Vedic religion of India is also a religion of revelation.

In the Judeo-Christian traditions the Scriptures are such a revelation. They are seen as God's words and descriptions of God's actions, committed to writing for the good of people. The Scriptures are God's revelation of Himself to His people.

There is some evidence from very ancient cultures and religions that there may well have been a primitive revelation at the dawn of history. This would explain why even the remotest tribes in all the continents have similar concepts of a 'High God', who is the Lord of all things and people and who claims man's obedience in social life and requires man's worship in both a social and a personal way as a sign of this obedience.

Now, in describing the God as known to ancient peoples, we have used metaphors – God is 'Lord', He must be 'obeyed', God 'reveals' His will to His people – we use human ideas to describe God.

Story and revelation
But it is more than that. The main way that people learn and write about religion is through story.

When the Hebrews wanted to write about their God, they wrote a 'story' of how God is 'Lord of all'.

When the Jews celebrate Passover, they tell the story of how God set them free from slavery in Egypt and chose them as a people.

When Islam wants to teach Muslims, they teach the Qur'an – God's story revealed through the Prophet.

When Christians tell of their religion, they always go back to the history of Jesus written by the gospel authors – the story of his life, death and resurrection and the meaning of that story.

When it comes to very difficult religious ideas, religions use other metaphors or imaginative comparisons. Like the scientist, who may hold sometimes that light is a wave and sometimes that it is a particle, so religion often holds together ideas that may seem opposite to each other.

Just as the scientist sees different theories as complementary rather than exclusive and contradictory, so beliefs can be complementary.

One such example is the Christian teaching that there is only one God, and yet in that one God there is Father, Son and Holy Spirit.

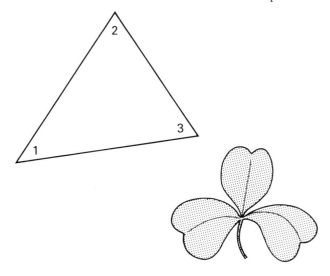

Figure 11 Three in one

are not three Gods but only one. Over the ages, Christians have used many ways of describing the Trinity, e.g. in pictures (Rublëv's icon of 'The Three Angels' (Figure 12)) or, in action words, God *sent* His Son, His Son *saves* people, and His Spirit *transforms* them; all such pictures are only partial, but taken together they give different insights, like a face seen through a shattered mirror.

This appears to be nonsense. How can three be one and one be three? Many examples have been used such as three angles in one triangle, or three leaves in one clover leaf (Figure 11).

But it still does not seem to make sense. If God exists, how can anyone possibly divide God up in this way?

The clue to the problem lies in seeing the difference between speaking about God as He is in Himself and speaking about God as Christians have come to know Him in their own experience.

1   The first Christians were all Jewish people who believed with great certainty that there is one God who is the Creator and Father of all.

2   But they had known Jesus and had become convinced, especially after his crucifixion and resurrection, that he was divine in a way in which people are not. Yet he prayed to God as his Father. So they called Jesus God's Son.

3   Jesus had promised to send the Holy Spirit to his disciples when he left them. The experience of the first Whitsun left a tremendous impression on the early Christians. They were sure that Jesus had indeed sent the Spirit. Yet the Spirit could not be totally different from Jesus, and Jesus could not be totally different from God, for there

Figure 12 Rublëv's icon of 'The Three Angels'

Of course in talking about God as Father, Son and Holy Spirit, language is stretched beyond the capacities of any human being to understand. Even greater imagination is called for here than in understanding, for example, the wave–particle dilemma. So for all of us in trying to understand anything about 'God' the voyage of discovery has really only just begun . . . .